ELECTROMAGNETIC WAVES
FOR **THERMONUCLEAR**
FUSION RESEARCH

ELECTROMAGNETIC WAVES
FOR **THERMONUCLEAR**
FUSION RESEARCH

Ernesto Mazzucato
Princeton Plasma Physics Laboratory, USA

World Scientific

NEW JERSEY • LONDON • SINGAPORE • BEIJING • SHANGHAI • HONG KONG • TAIPEI • CHENNAI

Published by

World Scientific Publishing Co. Pte. Ltd.

5 Toh Tuck Link, Singapore 596224

USA office: 27 Warren Street, Suite 401-402, Hackensack, NJ 07601

UK office: 57 Shelton Street, Covent Garden, London WC2H 9HE

Library of Congress Cataloging-in-Publication Data
Mazzucato, E. (Ernesto), author.
  Electromagnetic waves for thermonuclear fusion research / Ernesto Mazzucato (Princeton Plasma Physics Laboratory, USA).
    pages cm
  Includes bibliographical references and index.
  ISBN 978-9814571807 (hardcover : alk. paper)
  1. Controlled fusion. 2. Electromagnetic waves. I. Title.
  QC791.73.M39 2014
  621.48'4--dc23
                          2013047942

British Library Cataloguing-in-Publication Data
A catalogue record for this book is available from the British Library.

In-house Editor: Song Yu

Typeset by Stallion Press
Email: enquiries@stallionpress.com

Printed in Singapore

Front cover figure: Chaotic interference of waves reflected by a cutoff in the presence of a small level of short-scale turbulent fluctuations (Chapter 7).

The pure and simple truth is rarely pure and never simple

Oscar Wilde

To Alessandra

.

PREFACE

The science of magnetically confined plasmas covers a wide variety of topics, from classical and relativistic electrodynamics to quantum mechanics. During the last sixty years of research, our initial primitive understanding of plasma physics has made impressive progress thanks to a variety of experiments — from tabletop devices with plasma temperatures of a few thousands of degrees and confinement times of less than 100 microseconds, to large tokamaks with plasma temperatures of up to five hundred million degrees and confinement times approaching one second. We discovered that plasma confinement is impaired by a variety of instabilities leading to turbulent processes with scales ranging from the plasma size to a few millimeters. Understanding these phenomena, which are what has slowed down our progress towards a fusion reactor, has required the use of very sophisticated diagnostic tools, many of them employing electromagnetic waves

The primary objective of this book is to discuss the fundamental physics upon which the application of electromagnetic waves to the study of magnetically confined plasmas is based. My hope is to bring the reader who intends to enter the field of controlled thermonuclear fusion to a full knowledge of this type of diagnostics. Since the emphasis is on physics principles, the book does not provide a detailed description of techniques and instrumentation. Nevertheless, at the end of each chapter, a list of references should ease the acquisition of any technical information.

To help the reader to jump to his topic of interest, a short overview of the subject matter of the different chapters is as follows:

- Chapter 1 contains a brief summary of the state of the art of research on magnetically confined plasmas together with a description of the main features of a tokamak magnetic configuration.
- Chapter 2 is a review of the theory of propagation of electromagnetic waves with frequencies much larger than the ion cyclotron frequency in cold homogeneous plasmas.
- Chapter 3 reviews the theory of wave propagation in inhomogeneous plasmas and derives the ray equations of geometrical optics including first-order diffractive effects.
- Chapter 4 describes the methods of interferometry and polarimetry for the measurement of plasma density and magnetic field.
- Chapter 5 reviews the theory of collective scattering of electromagnetic waves by plasma density fluctuations and describes its application to the measurement of the short-scale turbulence that theory indicates as the major cause of anomalous transport in magnetically confined plasmas.
- Chapter 6 deals with the non-collective scattering of electromagnetic waves where the plasma behaves as an assembly of uncorrelated particles. The emphasis here is on the inclusion of relativistic effects in the theory of this phenomenon.
- Chapter 7 treats the subject of reflectometry, where a reflected wave from a plasma cutoff is used for inferring the plasma density and its fluctuations.
- Chapter 8 reviews the relativistic theory of electron cyclotron waves in magnetized plasmas.
- Chapter 9 describes the application of electron cyclotron emission to the measurement of the electron temperature and its fluctuations in magnetically confined plasmas.

The book could be used for a graduate course in plasma physics. Hopefully, it will contribute to the education of the next generation of fusion scientists, who — luckier than me — will have a chance of employing some of the methods described in this book for the study of plasmas in a real fusion reactor.

Ernesto Mazzucato
Princeton, NJ, USA

ABOUT THE AUTHOR

The author graduated from the University of Padua. After a brief tenure at the Laboratorio Gas Ionizzati of Frascati (Italy), he moved to the Princeton Plasma Physics Laboratory of Princeton University, where he developed some of the diagnostic methods described in this book and conducted numerous experiments on plasma confinement and turbulence in tokamaks. The author is a fellow of the American Physical Society.

CONTENTS

CHAPTER 1

CONTROLLED THERMONUCLEAR FUSION

The target of research on controlled thermonuclear fusion (CTF) is the conundrum of how to find a new source of energy capable of satisfying the needs of a growing world population without destroying the environment. The goal is to control the large amount of energy that is released when two light atomic nuclei fuse to form a heavier one.

In this book, the plasma theory behind some of the most important diagnostics used in thermonuclear fusion research on magnetically confined plasmas will be presented. Since most of these tools have been developed for diagnosing tokamak plasmas, we will often use examples taken from tokamak experiments for illustrating the problem under discussion.

In this chapter, the state of the art of CTF Research is briefly summarized together with a description of the main feature of the magnetic configuration of tokamaks.

1.1 Introduction

The basic concept behind any CTF approach is to bring the nuclear reactants to very high temperature and to keep the plasma energy confined for a time sufficiently long to allow a number of fusion reactions to take place — *sufficiently* meaning that the energy released to the plasma must be larger than the energy lost from thermal conduction and radiation losses, such as those from impurities, bremsstrahlung and electron cyclotron emission.

The type of useful fusion reactions must satisfy several criteria. First, they must be exothermic for obvious reasons. Second, they must involve only low-Z nuclei because otherwise the electrostatic repulsion would require enormous temperatures for their fusion. Third, they must require only two reactants since three-body collisions have very small cross-sections. Fourth, they must have at least two products to allow the simultaneous conservation of energy and momentum without relying on the electromagnetic force. At the present stage of CTF, of all the fusion reactions satisfying these criteria only those between the two isotopes of hydrogen — Deuterium $(D = {}^2_1H)$ and Tritium $(T = {}^3_1H)$ — are of any use. They are:

$$D + T \rightarrow {}^4He \,(3.5\,\text{Mev}) + n \,(14.1\,\text{Mev}),$$
$$D + D \rightarrow T \,(1.01\,\text{Mev}) + p \,(3.02\,\text{Mev}), \qquad (1.1)$$
$$D + D \rightarrow {}^3He \,(0.82\,\text{Mev}) + n \,(2.45\,\text{Mev}),$$

where p stands for proton and n for neutron, and brackets contain the energy of fusion products. The first of these reactions has the largest cross-section (Fig. 1.1) and will therefore be the first to be used in a fusion reactor. The last two occur with the same probability.

Deuterium is universally available in large quantities since about one part in 5000 of hydrogen in seawater is deuterium. As a potential

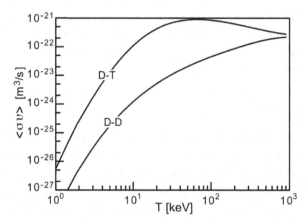

Figure 1.1.   Reactions rate $\langle \sigma v \rangle$ averaged over a Maxwellian distribution. The D–D curve is the sum of the rates of the last two fusion reactions in Eq. (1.1).

fuel of a fusion reactor, a liter of seawater could produce as much energy as 300 liters of oil. On the contrary, there is no sizable natural source of tritium because of its short radioactive half-life of 12.3 years. Thus, it must be obtained with breeding from lithium using the following reactions:

$$n + {}^6Li \rightarrow {}^4He + T + 4.8\,Mev,$$
$$n + {}^7Li \rightarrow {}^4He + T + n - 2.5\,Mev. \tag{1.2}$$

The first of these reactions is exothermic and therefore add a positive contribution to the reactor energy balance. The second is endothermic, but it has the advantage of conserving the number of neutrons. Lithium will be part of a blanket surrounding the plasma that will also contain a neutron multiplier (Beryllium or Lead) for obtaining a breeding ratio larger than one.

Like Deuterium, Lithium can be extracted from seawater where its average ion concentration is 0.17 ppm. Thus, these two elements as the fuel of a D–T fusion reactor could provide an effectively inexhaustible source of energy, lasting for more than a million years.

1.2 Ignition Conditions

By definition, a fusion reactor reaches ignition when plasma heating from the products of fusion reactions is sufficient for maintaining a constant plasma temperature without any external input of power. The Lawson criterion [1] sets the conditions for this to occur. Here we will repeat its derivation for the case of a D–T reactor (in SI units and temperatures in keV) but the same can be applied to other fusion fuels.

Let τ_E be the plasma energy confinement time as defined by the ratio of the total plasma energy E to the power loss P_L. Assuming the optimum 50/50 mixture of D–T, no impurities and the same temperature for ions and electrons, we get

$$E = 3n\kappa TV, \tag{1.3}$$

where n, T and V are, respectively, the plasma density, temperature and volume, and κ is the Boltzmann constant. In (1.3), both n and T

are assumed having flat spatial profiles. In the opposite case they should be replaced by proper averages.

The rate of fusion reactions is

$$F = n_D n_T \langle \sigma v \rangle V = \frac{1}{4} n^2 \langle \sigma v \rangle V, \tag{1.4}$$

so that under steady-state conditions we must have

$$P_L = \frac{E}{\tau_E} = \frac{1}{4} n^2 \langle \sigma v \rangle V E_F, \tag{1.5}$$

where E_F is the energy delivered to the plasma by fusion reactions (3.5 MeV for the first of (1.1)). From this we get

$$n\tau_E = \frac{12}{E_F} \frac{\kappa T}{\langle \sigma v \rangle}. \tag{1.6}$$

Since $T/\langle \sigma v \rangle$ has an absolute minimum at $T \approx 25\,\text{keV}$, (1.6) gives the Lawson criterion

$$n\tau_E > 1.5 \times 10^{20}\,\text{m}^{-3}\,\text{s}, \tag{1.7}$$

which must be satisfied in an ignited D–T plasma.

Since τ_E is itself a function of temperature, a better criterion is the triple product $nT\tau_E$, which can be obtained by multiplying both sides of (1.6) by T. To within 10%, $\langle \sigma v \rangle$ is proportional to T^2 in the temperature range 10–20 keV, which makes the triple product almost independent of T. Hence, in this range of temperatures the ignition condition becomes

$$nT\tau_E > 3 \times 10^{21}\,\text{m}^{-3}\,\text{keV s}. \tag{1.8}$$

The precise value of the constant in (1.8) depends on the profiles of density and temperature and on other factors such as the presence of impurities. For parabolic density and temperature profiles, the ignition condition becomes

$$n_0 T_0 \tau_E \geq 5 \times 10^{21}\,\text{m}^{-3}\,\text{keV s}, \tag{1.9}$$

where n_0 and T_0 are the peak values of n and T.

1.3 Tokamaks

Tokamaks belong to the group of magnetic confinement where, as the name indicates, magnetic fields are used for confining hot plasmas in a steady-state configuration. This is one of two major branches of research on fusion energy, the other being inertial confinement where the plasma is first compressed to very high densities, and then is allowed to expand freely once fusion reactions begin. Of the two approaches, magnetic confinement is more developed and is considered to be the first to lead to a fusion reactor.

In the early years of fusion research on magnetic confinement, a variety of magnetic schemes were investigated, from toroidal systems such as stellarators, tokamaks, reversed field pinches, to open systems such as mirrors and cusps. By the end of the nineties it became clear that tokamaks were by far the most successful (Fig. 1.2) and had the best chance of leading to a fusion reactor. This is why presently most of the research activity on magnetic confinement is on tokamaks, including the construction of the 500 MW prototype reactor — the International Thermonuclear Experimental Reactor (ITER) — that should become operational after 2025 [2].

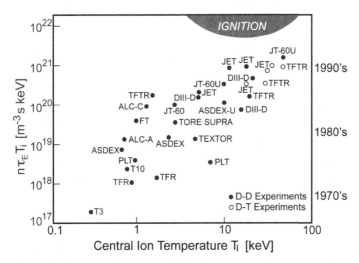

Figure 1.2. Triple products $n\tau_E T_i$ as a function of the central ion temperature for several tokamak experiments.

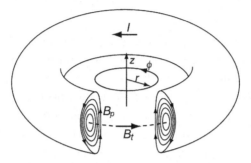

Figure 1.3. Schematic view of a tokamak toroidal configuration showing a crosscut of closed magnetic surfaces, with B_T and B_P the toroidal and poloidal magnetic fields and I the plasma toroidal current. The magnetic axis (dash line) is the magnetic line that comes back to the same position after one turn around the torus.

Tokamaks were developed in the Soviet Union from theoretical ideas of Tamm and Sakharov [3, 4]. They are characterized by an axially symmetric magnetic configuration (Fig. 1.3) where the plasma is confined by a strong toroidal magnetic field (B_T), which is produced mainly by external coils (toroidal magnet), together with the poloidal field (B_P) produced by a toroidal current flowing in the plasma and by external coils [5]. The field lines of the total magnetic field **B** produce a set of nested toroidal magnetic surfaces, as illustrated in Fig. 1.3. From the equilibrium equation (SI units)

$$\mathbf{j} \times \mathbf{B} = \nabla p, \qquad (1.10)$$

where **j** is the current density and p is the plasma pressure, we get $\mathbf{B} \cdot \nabla p = 0$. Thus, the surfaces of constant pressure coincide with the magnetic surfaces. From (1.10) we get also $\mathbf{j} \cdot \nabla p = 0$, and consequently the current lines lie on the magnetic surfaces as well.

To proceed, it is convenient to introduce the system of cylindrical coordinates (r, ϕ, z) of Fig. 1.3. From $\mathbf{B} = \nabla \times \mathbf{A}$, where **A** is the vector potential, we get

$$
\begin{aligned}
B_r &= -\frac{1}{r}\frac{\partial \psi}{\partial z}, \\
B_z &= \frac{1}{r}\frac{\partial \psi}{\partial r},
\end{aligned}
\qquad (1.11)
$$

with $\psi = rA_\phi$. From this it follows that $\mathbf{B} \cdot \nabla\psi = 0$, so that ψ is constant on the magnetic surfaces and thus may be used as their label. The poloidal flux through a circle C on the plane $z = 0$ with center on the axis of symmetry and radius r is

$$\int \mathbf{B} \cdot d\mathbf{S} = \int \nabla \times \mathbf{A} \cdot d\mathbf{S} = \int_C \mathbf{A} \cdot d\mathbf{l} = 2\pi r A_\phi = 2\pi\psi(r, 0),$$

showing that $2\pi\psi$ is the magnetic poloidal flux.

Similarly, from the Ampere's equation $\mu_0\mathbf{j} = \nabla \times \mathbf{B}$ we obtain

$$j_r = -\frac{1}{r}\frac{\partial J}{\partial z}, \quad j_z = \frac{1}{r}\frac{\partial J}{\partial r}, \tag{1.12}$$

where

$$J = \frac{rB_\phi}{\mu_0} \tag{1.13}$$

is the current poloidal flux per radian of ϕ, which is also constant on magnetic surfaces.

In conclusion, two of the three functions (ψ, J, p) can be considered a function of the third, as for instance

$$p = p(\psi), \quad J = J(\psi). \tag{1.14}$$

To describe the twisting of the magnetic field lines around the magnetic surfaces, it is useful to define the safety factor q as the limit

$$q = \lim_{N \to \infty} \frac{N_T}{N_P}, \tag{1.15}$$

where N_T and N_P are the numbers of toroidal and poloidal windings of a field line around the torus, and the limit is over an infinite number of revolutions. To demonstrate that this limit does not depend on the particular field line on a magnetic surface, let us consider the angular toroidal displacement $d\phi$ that a field line makes as it moves a distance dl in the poloidal plane, so that

$$\frac{rd\phi}{dl} = \frac{B_T}{B_P},$$

where B_T and B_P are the local toroidal and poloidal magnetic field, respectively. It follows that the total change in ϕ when the field

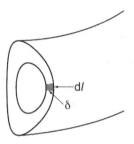

Figure 1.4.   Plasma layer between two neighboring flux surfaces.

line has completed a full poloidal revolution around the magnetic surface is

$$\Delta\phi = \oint \frac{B_T}{rB_P}dl, \tag{1.16}$$

where the integral is over a single poloidal revolution. Let us then consider the plasma layer between two neighboring magnetic surfaces, as illustrated in Fig. 1.4. The total poloidal flux through the layer is

$$d\Psi = 2\pi r B_P \delta, \tag{1.17}$$

(not to be confused with $d\psi = d\Psi/2\pi$) while the toroidal flux is

$$d\Phi = \oint (B_T \delta)dl. \tag{1.18}$$

Substituting (1.17) into (1.16) and using (1.18) gives

$$\Delta\phi = 2\pi \frac{d\Phi}{d\Psi}, \tag{1.19}$$

so that the safety factor q is given by

$$q = \frac{d\Phi}{d\Psi}, \tag{1.20}$$

showing that it is the rate of change of the toroidal flux with the poloidal flux, and is therefore another function of magnetic surfaces.

Substituting (1.11) and (1.12) into (1.10) yields

$$\nabla p = -\frac{B_\phi}{r}\nabla J + \frac{j_\phi}{r}\nabla\psi, \qquad (1.21)$$

which can be cast in the form

$$j_\phi = r\frac{dp}{d\psi} + B_\phi\frac{dJ}{d\psi}. \qquad (1.22)$$

This together with (1.13) leads to

$$j_\phi = r\frac{dp}{d\psi} + \frac{\mu_0}{r}J\frac{dJ}{d\psi}. \qquad (1.23)$$

Finally, from the ϕ component of the Ampere's equation together with (1.11) and (1.23), we obtain the Grad–Shafranov equation [6, 7]

$$r\frac{\partial}{\partial r}\left(\frac{1}{r}\frac{\partial\psi}{\partial r}\right) + \frac{\partial^2\psi}{\partial z^2} = -\mu_0 r^2\frac{dp}{d\psi} - \mu_0^2 J\frac{dJ}{d\psi} \qquad (1.24)$$

that together with (1.11) and (1.13), which may be written as

$$\mathbf{B} = \mu_0\frac{J}{r}\mathbf{e}_\phi + \frac{1}{r}(\nabla\psi \times \mathbf{e}_\phi), \qquad (1.25)$$

and the corresponding equation from (1.12) and (1.23)

$$\mathbf{j} = \left(r\frac{dp}{d\psi} + \mu_0\frac{J}{r}\frac{dJ}{d\psi}\right)\mathbf{e}_\phi + \frac{1}{r}(\nabla J \times \mathbf{e}_\phi) \qquad (1.26)$$

fully describe the tokamak equilibrium.

Given $p(\psi)$ and $J(\psi)$, a numerical solution of (1.24) yields $\psi(r, z)$. This can be done very easily when p and J take the simple form

$$p(\psi) = a\psi^2 \quad \text{and} \quad J^2(\psi) = J_0^2 + b\psi^2, \qquad (1.27)$$

with $\psi = 0$ at the plasma boundary, which allows to transform (1.24) into an ordinary linear differential equation. To see how this can be done [8], we look for solutions of (1.24) in the form

$$\psi = \sum_{i=1}^{n} f_i(r)[c_i\cos(k_i z) + s_i\sin(k_i z)], \qquad (1.28)$$

where c_i and s_i are constants and $f_i(r)$ are solutions of the ordinary differential equation

$$\frac{d^2 f_i}{dr^2} - \frac{1}{r}\frac{df_i}{dr} + \left(2\mu_0 a r^2 + \mu_0^2 b - k_i^2\right) f_i = 0. \qquad (1.29)$$

Although the change of independent variable $\rho = (a/2)^{1/2} r^2$ allows expressing the solution of this equation in terms of the Coulomb wave functions [9], it is easier to solve directly the ordinary differential equation with standard numerical methods. Accordingly, for given values of a and b, we first find n independent solutions of (1.29) satisfying the initial conditions $f_i = f_{i0}$ and $df_i/dr = d_i$ at $r = r_0$. Thus, inserting these into (1.28) gives ψ in terms of $5n + 3$ parameters. At this point, the problem has been reduced to that of finding the values of some of these parameters yielding the solution of (1.29) with closed magnetic surfaces and the desired shape. The physical dimensions of the magnetic configuration are then set by using the group of transformations $(r, z) \rightarrow (\alpha r, \alpha z)$, $k_i \rightarrow \alpha^{-1} k_i$, $a = \alpha^{-4} a$, $b = \alpha^{-2} b$, $f_i \rightarrow \alpha^2 f_i$ leaving (1.29) invariant. Finally, for a given vacuum toroidal magnetic field, the flux function ψ is renormalized to give a prescribed value of the safety factor $q = d\Phi/d\Psi$ (with the toroidal magnetic flux Φ given by (1.25)) at a certain location, usually the magnetic axis or the plasma edge. This last step yields the toroidal plasma current from (1.26).

Following this procedure, it is easy to produce a variety of tokamak configurations, as shown by Figs. 1.5 and 1.6, that display results obtained using only three terms in (1.28). In the following chapters, we will use very often this type of solutions of the Grad–Shafranov equation for illustrating the topic under discussion.

1.4 Tokamak Operating Limits

The ignition condition (1.8) can be cast in the form

$$B_T^2 \beta_T \tau_E \geq 2.4, \qquad (1.30)$$

with β_T the plasma beta as defined by the ratio of the volume average plasma pressure to the vacuum magnetic pressure at the

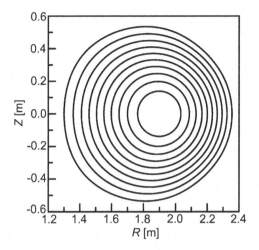

Figure 1.5. Poloidal cross-section of a circular shape tokamak.

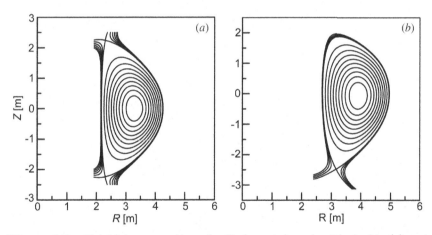

Figure 1.6. Poloidal cross-section of a D-shape tokamak with double (a) and single (b) null separatrix (the surface separating closed from open magnetic surfaces).

plasma center

$$\beta_T = \frac{2n\kappa T}{(B_T^2/2\mu_0)} = 8 \times 10^{-2}\frac{n_{20}T}{B_T^2}, \qquad (1.31)$$

where n_{20} is the plasma density in $10^{20}\,\mathrm{m}^{-3}$ and ions and electrons are assumed having equal temperatures (in keV). The terms on the left-hand side of (1.30) represent three of the major tokamak operating limits.

The magnetic field B_T is a technological problem of crucial importance since the fusion power scales like B_T^4. In present tokamaks using copper coils, the toroidal magnet field in the plasma center can be as high as $13\,\mathrm{T}$ [10], the limit being set by magnetic forces and the cooling of coils. However, a steady-state reactor will be forced to use superconducting coils since the Joule heating losses in copper coils will be unacceptable. In this case the limitation will be set by the loss of superconductivity above a critical magnetic field. Present technology seems to limit the maximum field at the coil conductor to about $12\,\mathrm{T}$ [2], which for the magnet of a tokamak reactor would limit the magnetic field in the plasma center to about 5–6 T.

The plasma beta is also an important parameter since the fusion power scales like β_T^2. In present day tokamaks, it is limited by a variety of magnetohydrodynamic (MHD) instabilities [11, 12] to a value of

$$\beta_T = 1 \times 10^{-2} \beta_N \frac{I}{aB_T}, \qquad (1.32)$$

where I is the total toroidal current in MA, a is the plasma minor radius in meters (Fig. 1.7) and the coefficient β_N is in the range

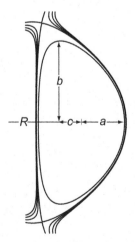

Figure 1.7.   Tokamak poloidal cross-section showing the 95% flux surface with elongation $\varepsilon = b/a$ and triangularity $\delta = c/a$.

from 2 to 3. MHD plasma stability also limits the safety factor at the plasma edge to values larger than ~3. Thus, since the safety factor at the 95% flux surface is [12]

$$q_{95} \approx 5 \frac{aB_T}{I} \frac{a}{R} \frac{1 + \varepsilon^2(1 + 2\delta^2)}{2}, \tag{1.33}$$

where R is the major radius of the plasma torus with elongation $\varepsilon = b/a$ and triangularity $\delta = c/a$ (Fig. 1.7), the restriction on the minimum value of q_{95} implies one on the maximum value of I/aB_T and hence on the maximum value of β_T. Equation (1.33) shows also the beneficial effects of elongation and triangularity. In present day tokamaks with $R/a \sim 3\text{--}4$, $\varepsilon \sim 1.8$ and $\delta \sim 0.5$, β_T is in the range of a few percent.

Another restriction is imposed by the density limit [13]

$$n < \frac{I}{\pi a^2}, \tag{1.34}$$

again in units of $10^{20}\,\mathrm{m}^{-3}$ and I in MA, which seems to be caused by an increase in radiation losses from impurities as the density rises. Since this radiation is predominantly from the plasma edge, the resulting plasma cooling produces a contraction of the current channel with a decrease in q, making the plasma highly unstable.

Finally, τ_E is another parameter of vital importance for the realization of an economically feasible fusion reactor, since the worse the energy confinement, the larger must be the reactor. Unfortunately, at present there is not a unified theory of plasma transport in tokamaks. What is known is that the transport observed in tokamak experiments exceeds the predictions of classical collisional theory by several orders of magnitude. The plasma physics literature abounds with explanations of this important phenomenon [14–17] — all invoking some kind of turbulent process — but none is based on a self-consistent theory. The difficulty stems from the fact that the turbulence of magnetically confined plasmas is not only a highly nonlinear phenomenon, as in fluid hydrodynamics, but it can be driven by a large variety of instabilities as well. The second problem is that it is very difficult if not impossible to perform a comprehensive set of turbulence measurements in the inhospitable environment of hot plasmas. Inevitably, one is forced to rely on scaling laws derived from experiments and dimensional analysis.

In a fully developed turbulence the relevant parameters for determining the induced transport are the turbulence scale and time lengths, i.e., the correlation time and size of turbulent structures. If we indicate with τ and Δ these quantities and assume a random walk model for transport, we expect the heat conductivity to be of the order of $\chi \approx \Delta^2/\tau$. In confined magnetized plasmas, a variety of waves are driven unstable by density and temperature gradients. The characteristic frequency of these instabilities is the drift frequency

$$\omega^* = k_\perp \frac{\kappa T}{eB} \frac{\nabla n}{n} \approx k_\perp \frac{\kappa T}{eBa}, \qquad (1.35)$$

where $k_\perp$ is the wave number perpendicularly to $\boldsymbol{B}$ and $\nabla n \approx n/a$. Then, if we take $\Delta \approx k_\perp^{-1}$ and $\tau \approx 1/\omega^*$, the ansatz for the heat conductivity is

$$\chi \approx \frac{1}{k_\perp a} \frac{\kappa T}{eB}. \qquad (1.36)$$

In the worst-case scenario where $k_\perp \approx 1/a$, this becomes the Bohm diffusion coefficient that early studies found in the Model C-Stellarator [18]. The other extreme case has $k_\perp \approx 1/\rho_i$ (where $\rho_i = v_{ti}/\omega_i$ is the ion Larmor radius, with $v_{ti} = (\kappa T_i/m_i)^{1/2}$ the ion thermal velocity and $\omega_i = e_i B/m_i$ the ion cyclotron frequency), which is the characteristic wave number of the Ion Temperature Gradient Mode [14–17]. It leads to the Gyro–Bohm scaling

$$\chi \approx \frac{\rho_i}{a} \frac{kT}{eB} \qquad (1.37)$$

and the energy confinement time

$$\tau_E \propto \frac{a^2}{\chi} \propto \frac{1}{B} \frac{a^3}{\rho_i^3}. \qquad (1.38)$$

At constant values of q_{95}, ε and δ, this together with (1.32) and (1.34) (written as $n = n_N I/\pi a^2$) yield

$$nT\tau_E \propto \frac{n_N^{3/2}}{\beta_N^{1/2}} \frac{(Ba)^{5/2}}{R}. \qquad (1.39)$$

One of the most exiting results from the last years of tokamak experiments has been the discovery of a self-organization of turbulence, where the nonlinear interaction of turbulent modes causes the plasma to rotate with a shear velocity that decorrelates and breaks the turbulent eddies. One of these phenomena is the H-mode — discovered in the ASDEX tokamak [19] — where the onset of a rotating plasma layer at the plasma boundary causes a decrease in the level of turbulence and a rise in the energy confinement by a factor of two. The H-mode has been observed in many tokamak experiments following the scaling [20]

$$\tau_E B \propto (\rho_i/a)^{-2.70} \beta^{-0.90} (R/a)^{-0.73}, \tag{1.40}$$

leading to

$$n T \tau_E \propto \frac{n_N^{1.32}}{\beta_N^{1.32}} \frac{B^{2.35} a^{2.21}}{R^{0.85}}, \tag{1.41}$$

which shows a remarkable similarity to (1.39).

In conclusion, research on tokamaks has made great progress, bringing this confinement scheme close to ignition (Fig. 1.2). In this book, we will discuss some of the diagnostic tools that made it possible.

Bibliography

[1] Lawson, J. D., *Proc. Phys. Soc. B*, **70**, 6 (1957).
[2] ITER Technical Basis, *ITER EDA Documentation Series No. 24*, IAEA, Vienna, 2002.
[3] Tamm, I. E., in *Plasma Physics and the Problem of Controlled Thermonuclear Reactions*, Vol. 1, Edited by Leontovich M. A., Pergamon Press, New York, 1962, p. 1.
[4] Sakharov, A. D., in *Plasma Physics and the Problem of Controlled Thermonuclear Reactions*, Vol. 1, Edited by Leontovich M. A., Pergamon Press, New York, 1962, p. 21.
[5] Solovev, L. S. and Shafranov, V. D., in *Review of Plasma Physics*, Vol. 5, Edited by Leontovich, M. A., Consultants Bureau, New York, 1966, p. 1.
[6] Grad, H. and Rubin, H., in *Proc. 2nd United Nations Int. Conf. Peaceful Uses of Atomic Energy*, United Nations, Geneva, 1958.
[7] Shafranov, V. D., *Sov. Phys. JETP* **6**, 545 (1958).
[8] Mazzucato, E., *Phys. Plasmas* **3**, 441 (1996).
[9] Abramowitz, M. and Stegun, A., in *Handbook of Mathematical Functions*, Dover, New York, 1968, p. 537.

[10] Coppi, B., Nassi, M. and Sugiyama, L. E., *Phys. Scr.* **45**, 112 (1992).

[11] Wesson, J., *Tokamaks*, Clarendon Press, Oxford, 1997.

[12] Freiberg, J. P., *Plasma Physics and Fusion Energy*, Cambridge University Press, Cambridge, 2007.

[13] Greenwald, M., *Plasma Phys. Control. Fusion* **44**, R27 (2002).

[14] Coppi, B. and Rewoldt, *Advances in Plasma Physics*, Vol. 6, Edited by Simon, A. and Thompson, W. B., Wiley, New York, 1976, p. 421.

[15] Connor, J. W. and Wilson H. R., *Plasma Phys. Control. Fusion* **36**, 719 (1994).

[16] Horton, W., *Rev. Mod. Phys.* **71**, 735 (1999).

[17] Horton, W., *Turbulent Transport in Magnetized Plasmas*, World Scientific, London, 2012.

[18] Bishop, A. S. and Hinnov, E., in *Plasma Physics and Controlled Nuclear Fusion Research*, Vol. 2, Vienna, IAEA, 1966, p. 673.

[19] Wagner, F., Becker, G., Behringer, K., Campbell, D. *et al.*, *Phys. Rev. Lett.* **49**, 1408 (1982).

[20] ITER Physics Expert Groups On Confinement and Transport, *Nucl. Fusion* **39**, 2175 (1999).

CHAPTER 2

ELECTRON WAVES

In most applications of electromagnetic waves to plasma diagnostics, it is sufficient to consider waves propagating in cold plasmas with frequency much larger than the ion cyclotron frequency, i.e., neglecting the ion dynamics. In this chapter, we will review the theory of propagation of these waves in homogeneous plasmas.

2.1 Maxwell Equations

The propagation of electromagnetic waves is governed by Maxwell equations that in Gaussian units are

$$\nabla \times \mathbf{E} + \frac{1}{c}\frac{\partial \mathbf{B}}{\partial t} = 0, \tag{2.1}$$

$$\nabla \times \mathbf{B} - \frac{1}{c}\frac{\partial \mathbf{E}}{\partial t} = \frac{4\pi}{c}\mathbf{j}, \tag{2.2}$$

$$\nabla \cdot \mathbf{B} = 0, \tag{2.3}$$

$$\nabla \cdot \mathbf{E} = 4\pi\rho, \tag{2.4}$$

where $\mathbf{B}$ and $\mathbf{E}$ are the magnetic and electric fields, and $\mathbf{j}$ and ρ are the current and space charge densities induced by the wave.

The Lorentz force on a test charge q moving with velocity $\mathbf{v}$

$$\mathbf{F} = q\left(\mathbf{E} + \frac{\mathbf{v}}{c} \times \mathbf{B}\right), \tag{2.5}$$

defines the physical meaning of $\mathbf{E}$ and $\mathbf{B}$. Divergence of (2.2) and
the time derivative of (2.4) yields the conservation of charge

$$\frac{\partial \rho}{\partial t} + \nabla \cdot \mathbf{j} = 0. \tag{2.6}$$

From (2.1) and (2.2), we get the equation involving only $\mathbf{E}$ and $\mathbf{J}$

$$\nabla \times \nabla \times \mathbf{E} + \frac{1}{c^2}\frac{\partial^2 \mathbf{E}}{\partial t^2} = -\frac{4\pi}{c^2}\frac{\partial \mathbf{J}}{\partial t}. \tag{2.7}$$

It is convenient to introduce the electric displacement vector

$$\mathbf{D}(t, \mathbf{r}) = \mathbf{E}(t, \mathbf{r}) + 4\pi \int_{-\infty}^{t} \mathbf{j}(t', \mathbf{r})dt' \tag{2.8}$$

and write (2.2) as

$$\nabla \times \mathbf{B} = \frac{1}{c}\frac{\partial \mathbf{D}}{\partial t}. \tag{2.9}$$

To close this system of equations, we need to know the dependence
on $\mathbf{E}$ of the induced currents and the electric displacement. For a
linear medium, the most general assumptions are [1]

$$\mathbf{j}(t, \mathbf{r}) = \int_{-\infty}^{t} dt' \int \boldsymbol{\sigma}(t, t', \mathbf{r}, \mathbf{r}') \cdot \mathbf{E}(t', \mathbf{r}')d\mathbf{r}', \tag{2.10}$$

$$\mathbf{D}(t, \mathbf{r}) = \int_{-\infty}^{t} dt' \int \boldsymbol{\varepsilon}(t, t', \mathbf{r}, \mathbf{r}') \cdot \mathbf{E}(t', \mathbf{r}')d\mathbf{r}', \tag{2.11}$$

where $\boldsymbol{\sigma}$ and $\boldsymbol{\varepsilon}$ are the electric conductivity and the dielectric tensors.
These equations, where the assumption is that the field is applied
at $t = -\infty$ and the effects of initial conditions are negligible, are a
consequence of the principle of causality stating that currents and
charges at time t are only caused by the electromagnetic field at
earlier times.

2.2   Homogeneous Plasmas

For a plasma that is homogeneous in both space and time, the kernels
of equations (2.10) and (2.11) must be functions only of $t - t'$ and

$\mathbf{r} - \mathbf{r}'$, so that we can write

$$\mathbf{j}(t, \mathbf{r}) = \int_{-\infty}^{t} dt' \int \boldsymbol{\sigma}(t - t', \mathbf{r} - \mathbf{r}') \cdot \mathbf{E}(t', \mathbf{r}')d\mathbf{r}', \qquad (2.12)$$

$$\mathbf{D}(t, \mathbf{r}) = \int_{-\infty}^{t} dt' \int \boldsymbol{\varepsilon}(t - t', \mathbf{r} - \mathbf{r}') \cdot \mathbf{E}(t', \mathbf{r}')d\mathbf{r}'. \qquad (2.13)$$

Let us now introduce the Fourier–Laplace transform of the field quantities, which for $\mathbf{E}(t, \mathbf{r})$ is

$$\mathbf{E}(\omega, \mathbf{k}) = \int_{0}^{\infty} dt \int \mathbf{E}(t, \mathbf{r}) \exp(-i\mathbf{k} \cdot \mathbf{r} + i\omega t)d\mathbf{r}, \qquad (2.14)$$

with its inverse

$$\mathbf{E}(t, \mathbf{r}) = \frac{1}{(2\pi)^4} \int_{C} d\omega \int \mathbf{E}(\omega, \mathbf{k}) \exp(i\mathbf{k} \cdot \mathbf{r} - i\omega t)d\mathbf{k}, \qquad (2.15)$$

where C is the Laplace contour parallel to the real ω axis and above all singularities.

Since (2.12) is a convolution integral, we have

$$\mathbf{j}(\omega, \mathbf{k}) = \boldsymbol{\sigma}(\omega, \mathbf{k}) \cdot \mathbf{E}(\omega, \mathbf{k}), \qquad (2.16)$$

where

$$\boldsymbol{\sigma}(\omega, \mathbf{k}) = \int_{0}^{\infty} dt \int \boldsymbol{\sigma}(t, \mathbf{r}) \exp(-i\mathbf{k} \cdot \mathbf{r} + i\omega t)d\mathbf{r}. \qquad (2.17)$$

Similarly, from (2.13)

$$\mathbf{D}(\omega, \mathbf{k}) = \boldsymbol{\varepsilon}(\omega, \mathbf{k}) \cdot \mathbf{E}(\omega, \mathbf{k}), \qquad (2.18)$$

where

$$\boldsymbol{\varepsilon}(\omega, \mathbf{k}) = \int_{0}^{\infty} dt \int \boldsymbol{\varepsilon}(t, \mathbf{r}) \exp(-i\mathbf{k} \cdot \mathbf{r} + i\omega t)d\mathbf{r}. \qquad (2.19)$$

For plane waves in a uniform plasmas, Maxwell equations (2.1) and (2.2) become

$$\mathbf{k} \times \mathbf{E}(\omega, \mathbf{k}) = \frac{\omega}{c}\mathbf{B}(\omega, \mathbf{k}), \qquad (2.20)$$

$$\mathbf{k} \times \mathbf{B}(\omega, \mathbf{k}) = -\frac{\omega}{c}\boldsymbol{\varepsilon}(\omega, \mathbf{k}) \cdot \mathbf{E}(\omega, \mathbf{k}), \qquad (2.21)$$

with

$$\boldsymbol{\varepsilon}(\omega, \mathbf{k}) = \mathbf{I} + \frac{4\pi i}{\omega}\boldsymbol{\sigma}(\omega, \mathbf{k}), \tag{2.22}$$

where $\mathbf{I}$ the unit tensor.

Since $\boldsymbol{\varepsilon}(t, \mathbf{r})$ is a real variable, we must have

$$\boldsymbol{\varepsilon}(\omega, \mathbf{k}) = \boldsymbol{\varepsilon}^*(-\omega^*, -\mathbf{k}^*), \tag{2.23}$$

where the asterisk indicates the complex conjugate value.

The components of the dielectric tensor $\boldsymbol{\varepsilon}$ are not all independent. In fact, from very general irreversible thermodynamic principles — the Onsager reciprocal relations [2] — it is possible to show that for any non-active medium in an external magnetic field $\mathbf{B}_0$, we must have

$$\varepsilon_{ij}(\omega, \mathbf{k}, \mathbf{B}_0) = \varepsilon_{ji}(\omega, -\mathbf{k}, -\mathbf{B}_0). \tag{2.24}$$

The symmetry implied by this relation can be made explicit by using the Cartesian coordinate system (x_1, x_2, x_3) of Fig. 2.1, in which $\mathbf{B}_0$ is along the x_3 direction and $\mathbf{k}$ is in the (x_1, x_3) plane [3].

Reversing the directions of $\mathbf{B}_0$ and $\mathbf{k}$ is equivalent to reversing the axes x_1 and x_3, i.e., performing a rotation of the reference frame

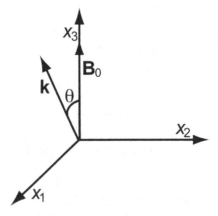

Figure 2.1.   Cartesian coordinate system with magnetic field $\mathbf{B}_0$ along the x_3 axis and wave vector $\mathbf{k}$ in the (x_1, x_3) plane.

by 180° about the x_2 axis. Since the matrix of this rotation is

$$\mathbf{R} = \begin{pmatrix} -1 & 0 & 0 \\ 0 & 1 & 0 \\ 0 & 0 & -1 \end{pmatrix}, \qquad (2.25)$$

the Onsanger relation can be written as

$$\mathbf{R} \cdot \boldsymbol{\varepsilon} \cdot \mathbf{R}^{\mathrm{T}} = \boldsymbol{\varepsilon}^{\mathrm{T}},$$

where the superscript T indicates the transpose. Hence, in general, the tensor $\boldsymbol{\varepsilon}$ has only six independent components and must have the form

$$\boldsymbol{\varepsilon}(\omega, \mathbf{k}, \mathbf{B}_0) = \begin{pmatrix} \varepsilon_{11} & \varepsilon_{12} & \varepsilon_{13} \\ -\varepsilon_{12} & \varepsilon_{22} & \varepsilon_{23} \\ \varepsilon_{13} & -\varepsilon_{23} & \varepsilon_{33} \end{pmatrix}. \qquad (2.26)$$

Note that the symmetry of (2.26) is valid only for the reference frame of Fig. 2.1; for a general frame, only the Onsager symmetry (2.24) holds.

If the medium is not spatially dispersive, i.e., if $\boldsymbol{\varepsilon}$ does not depend on $\mathbf{k}$, the dielectric tensor must be invariant under an arbitrary rotation (θ) around the x_3 axis

$$\mathbf{R}(\theta) \cdot \boldsymbol{\varepsilon} \cdot \mathbf{R}^{\mathrm{T}}(\theta) = \boldsymbol{\varepsilon}, \qquad (2.27)$$

with the rotation matrix

$$\mathbf{R}(\theta) = \begin{pmatrix} \cos\theta & \sin\theta & 0 \\ -\sin\theta & \cos\theta & 0 \\ 0 & 0 & 1 \end{pmatrix}, \qquad (2.28)$$

giving

$$\boldsymbol{\varepsilon}(\omega, \mathbf{B}_0) = \begin{pmatrix} \varepsilon_{11} & \varepsilon_{12} & 0 \\ -\varepsilon_{12} & \varepsilon_{11} & 0 \\ 0 & 0 & \varepsilon_{33} \end{pmatrix}. \qquad (2.29)$$

Finally, if $\mathbf{B}_0 = 0$, $\boldsymbol{\varepsilon}$ must be invariant and symmetric under rotations about $\mathbf{k}$. Taking the latter along the x_3 axis, we get

$$\boldsymbol{\varepsilon}(\omega, \mathbf{k}) = \begin{pmatrix} \varepsilon_{11} & 0 & 0 \\ 0 & \varepsilon_{11} & 0 \\ 0 & 0 & \varepsilon_{33} \end{pmatrix}. \tag{2.30}$$

From (2.7), we obtain the wave equation

$$\mathbf{k} \times (\mathbf{k} \times \mathbf{E}(\omega, \mathbf{k})) + \frac{\omega^2}{c^2} \boldsymbol{\varepsilon} \cdot \mathbf{E}(\omega, \mathbf{k}) = 0, \tag{2.31}$$

which represents a set of homogeneous algebraic equations, whose condition for non-trivial solutions gives the dependence of ω on $\mathbf{k}$, i.e., the wave dispersion relation [3–5]

$$D(\omega, \mathbf{N}) = 0, \tag{2.32}$$

where we have introduced the index of refraction $\mathbf{N} = \mathbf{k}c/\omega$.

Finally, in a plasma with weak wave absorption, for which the wave is essentially monochromatic, the average power density dissipation is

$$Q = \frac{1}{4}(\mathbf{j} \cdot \mathbf{E}^* + \mathbf{j}^* \cdot \mathbf{E}). \tag{2.33}$$

From this, (2.16) and (2.22), we get (with the usual convention on summation of repeated indexes)

$$Q = \frac{1}{4}(\sigma_{ij} E_j E_i^* + \sigma_{ij}^* E_j^* E_i) = \frac{\omega}{8\pi i} \frac{\varepsilon_{ij} - \varepsilon_{ji}^*}{2} E_i^* E_j, \tag{2.34}$$

from which we conclude that wave absorption depends only on the anti-Hermitian part of the dielectric tensor.

2.3   Plane Waves in Cold Homogeneous Plasmas

The wave dispersion relation in cold homogeneous plasmas can be derived from the equation of motion of a single electron

$$-i\omega\, m_e \mathbf{v} = -e\left(\mathbf{E} + \frac{\mathbf{v}}{c} \times \mathbf{B}_0\right), \tag{2.35}$$

where $\mathbf{B}_0$ is a constant magnetic field. In this equation, all first-order quantities are assumed to vary as $\exp i(\mathbf{k} \cdot \mathbf{r} - \omega t)$ and higher order terms are neglected. Then, the three components of (2.35) are

$$-i\omega m_e v_1 = -eE_1 - \frac{eB_0 v_2}{c},$$

$$-i\omega m_e v_2 = -eE_2 + \frac{eB_0 v_1}{c}, \qquad (2.36)$$

$$-i\omega m_e v_3 = -eE_3,$$

from which we readily get

$$v_1 = -i\frac{c}{B_0}\frac{\omega_c}{\omega}\frac{1}{1 - \frac{\omega_c^2}{\omega^2}}\left(E_1 - i\frac{\omega_c}{\omega}E_2\right),$$

$$v_2 = -i\frac{c}{B_0}\frac{\omega_c}{\omega}\frac{1}{1 - \frac{\omega_c^2}{\omega^2}}\left(E_2 + i\frac{\omega_c}{\omega}E_1\right), \qquad (2.37)$$

$$v_3 = -i\frac{c}{B_0}\frac{\omega_c}{\omega}E_3,$$

where $\omega_c = eB_0/m_e c$ is the electron cyclotron frequency. Inserting these into the conductivity equation

$$\mathbf{j} \equiv -n_e e\mathbf{v} = \boldsymbol{\sigma} \cdot \mathbf{E}, \qquad (2.38)$$

where n_e is the electron density, we get the conductivity tensor and from (2.22) the dielectric tensor

$$\boldsymbol{\varepsilon} = \begin{pmatrix} \varepsilon_{11} & \varepsilon_{12} & 0 \\ -\varepsilon_{12} & \varepsilon_{11} & 0 \\ 0 & 0 & \varepsilon_{33} \end{pmatrix}, \qquad (2.39)$$

with

$$\varepsilon_{11} = 1 - \frac{X}{1 - Y^2}, \quad \varepsilon_{12} = i\frac{XY}{1 - Y^2}, \quad \varepsilon_{33} = 1 - X,$$

where $X = (\omega_p/\omega)^2$, $\omega_p = (4\pi n_e e^2/m_e)^{1/2}$ (the electron plasma frequency) and $Y = \omega_c/\omega$. Inserting (2.39) into (2.31) yields the equation

$$
\begin{pmatrix}
\varepsilon_{11} - N_{\parallel}^2 & \varepsilon_{12} & N_{\parallel}N_{\perp} \\
-\varepsilon_{12} & \varepsilon_{11} - N^2 & 0 \\
N_{\parallel}N_{\perp} & 0 & \varepsilon_{33} - N_{\perp}^2
\end{pmatrix}
\begin{pmatrix}
E_2 \\
E_2 \\
E_3
\end{pmatrix} = 0,
\qquad (2.40)
$$

where $\mathbf{N} = N_{\perp}\mathbf{e}_1 + N_{\parallel}\mathbf{e}_3$. The condition for the existence of non-trivial solutions provides the dispersion relation

$$
\begin{aligned}
D(\mathbf{N},\omega) &\equiv \varepsilon_{11}N_{\perp}^4 - N_{\perp}^2[\varepsilon_{12}^2 + (\varepsilon_{11} + \varepsilon_{33})(\varepsilon_{11} - N_{\parallel}^2)] \\
&\quad + \varepsilon_{33}[(\varepsilon_{11} - N_{\parallel}^2)^2 + \varepsilon_{12}^2] = 0,
\end{aligned}
\qquad (2.41)
$$

from which we get two characteristic modes of propagation with refractive indexes

$$
N_{\perp}^2 = 1 - N_{\parallel}^2 - X \pm \frac{XY}{2}\frac{\Delta - Y(1 - N_{\parallel}^2)}{1 - X - Y^2},
\qquad (2.42)
$$

where $\Delta = [(1 - N_{\parallel}^2)^2 Y^2 + 4N_{\parallel}^2(1 - X)]^{1/2}$. If θ is the angle between $\mathbf{N}$ and $\mathbf{B}_0$, $N_{\parallel} = N\cos\theta$ and $N_{\perp} = N\sin\theta$, which can be cast in the form (the Appleton–Hartree formula [4])

$$
N^2 = 1 - \frac{2X(1 - X)}{2(1 - X) - Y^2\sin^2\theta \pm \Gamma},
\qquad (2.43)
$$

with $\Gamma = [Y^4\sin^4\theta + 4Y^2(1 - X)^2\cos^2\theta]^{1/2}$. The solution with the $+$ sign in front of Γ goes under the name of the *ordinary mode* (N_O) of propagation, while the other is the *extraordinary mode* (N_X).

From (2.41), we get that one (and only one) value of $N_{\perp}^2$ is zero when one of the two equations

$$
\varepsilon_{33} = 0,
$$

$$
(\varepsilon_{11} - N_{\parallel}^2)^2 + \varepsilon_{12}^2 = 0,
$$

is satisfied, i.e., when

$$
\begin{aligned}
X &= 1, \\
X &= (1 - N_{\parallel}^2)(1 \pm Y),
\end{aligned}
\qquad (2.44)
$$

and thus $N_O = 0$ for $X = 1$, and $N_X = 0$ for $X = 1 \pm Y$. These are referred to as the O, R, and L cutoff conditions, respectively, with the corresponding frequencies given by

$$\omega_O = \omega_p,$$

$$\omega_R = \left(\frac{\omega_c^2}{4} + \omega_p^2\right)^{1/2} + \frac{\omega_c}{2},$$ (2.45)

$$\omega_L = \left(\frac{\omega_c^2}{4} + \omega_p^2\right)^{1/2} - \frac{\omega_c}{2},$$

which satisfy the equations $\omega_L < \omega_O < \omega_R$ and $\omega_R > \omega_c$. Finally, from (2.43) we get that $N^2 \Rightarrow \infty$ for one of the two modes when

$$X = \frac{1 - Y^2}{1 - Y^2 \cos^2 \theta},$$ (2.46)

which for $\theta = \pi/2$ occurs at the upper-hybrid resonance frequency

$$\omega_{UH} = (\omega_c^2 + \omega_p^2)^{1/2}.$$ (2.47)

For wave propagation parallel to $\mathbf{B}_0$ ($\theta = 0, N_\perp = 0$), (2.43) yields

$$N^2 = 1 - \frac{X}{1 \pm Y},$$ (2.48)

with

$$\frac{E_1}{E_2} = \pm i, \quad E_3 = 0,$$ (2.49)

making the wave polarization circular since the two non-zero electric components are in quadrature and have the same amplitude. For the plus sign, an observer looking along the direction of wave propagation would see the electric field rotating anticlockwise (left-handed polarization). For the other case, the rotation would be clockwise (right-handed polarization).

For wave propagation perpendicular to $\mathbf{B}_0 (\theta = \pi/2, N_\parallel = 0)$, the solutions are

$$N_O^2 = 1 - X,$$
$$N_X^2 = 1 - \frac{X(1-X)}{1 - X - Y^2}, \qquad (2.50)$$

with wave polarization

$$E_1 = E_2 = 0, \qquad (2.51)$$

and

$$\frac{E_1}{E_2} = -i\frac{1 - X - Y^2}{XY}, \quad E_3 = 0. \qquad (2.52)$$

The first case has a linear polarization with $\mathbf{E}$ parallel to $\mathbf{B}_0$, while the second has an elliptical polarization with $\mathbf{E}$ perpendicular to $\mathbf{B}_0$.

For $Y < 1$, which is the case of interest for plasma diagnostics, Fig. 2.2 shows how the dependence of N^2 on the value of X changes as the direction of wave propagation goes from transverse to longitudinal with respect to $\mathbf{B}_0$. In these plots, where the modes can be identified using the cutoff conditions (2.44), we see that as the angle with the magnetic field decreases, the refractive index of the two modes undergoes a continuous deformation before finally coalescing at $\theta = 0°$ and $X = 1$, which is when both numerator and denominator in (2.43) become equal to zero. The same occurs for $Y > 1$ (Fig. 2.3), the only difference being that in this case, it is the refractive index of the O-mode that has an asymptote for small values of θ and positive values of X.

These figures show how even such an apparently simple phenomenon as the propagation of electromagnetic waves in a cold homogeneous plasma has some intricate features. This explains the difficulty of fully understanding some of the much more complex physical processes that are taking place in hot non-homogeneous plasmas.

2.4   Wave Polarization

In a following chapter, we will need the wave polarization on the plane perpendicular to $\mathbf{k}$. For this, let us change the system of coordinates

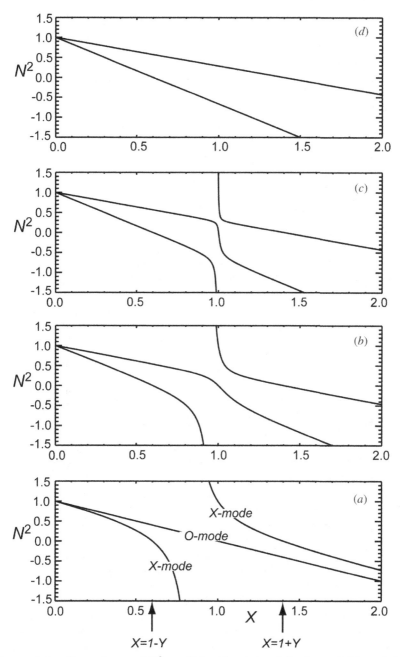

Figure 2.2. Dependence of N^2 on X for $Y = 0.4$ and for $\theta = 90°$ (a), $30°$ (b), $10°$ (c), and $0°$ (d).

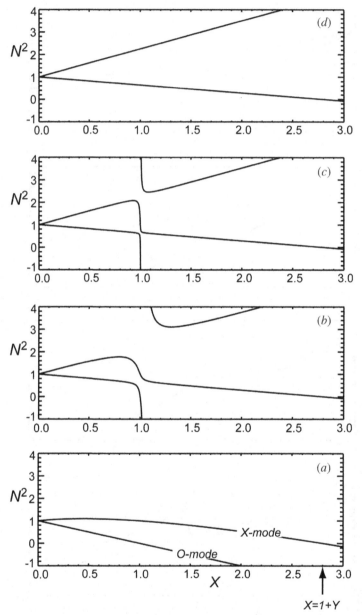

Figure 2.3. Dependence of N^2 on X for $Y = 1.8$ and for $\theta = 90°$ (a), $10°$ (b), $3°$ (c), and $0°$ (d).

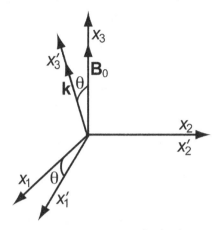

Figure 2.4. Cartesian coordinate system (x'_1, x'_2, x'_3) with the x'_3 axis along **k** obtained with a rotation of the frame of Fig. 2.1 around the x_2 axis.

(x_1, x_2, x_3) of Fig. 2.1, used previously for obtaining the dielectric tensor (2.39), to the system (x'_1, x'_2, x'_3) of Fig. 2.4 where the wave vector **k** is parallel to the x'_3 axis and the magnetic field $\mathbf{B}_0$ is in the (x'_1, x'_3) plane. This is obtained by rotating the frame of Fig. 2.1 by an angle θ around the x_2 axis.

The corresponding rotation matrix is

$$\mathbf{R} = \begin{pmatrix} \cos\theta & 0 & -\sin\theta \\ 0 & 1 & 0 \\ \sin\theta & 0 & \cos\theta \end{pmatrix}, \tag{2.53}$$

which from (2.39) yields

$$\boldsymbol{\varepsilon}' = \begin{pmatrix} 1 - \dfrac{X(1 - Y^2 \sin^2\theta)}{1 - Y^2} & i\cos\theta\dfrac{XY}{1 - Y^2} & -\sin\theta\cos\theta\dfrac{XY^2}{1 - Y^2} \\ -i\cos\theta\dfrac{XY}{1 - Y^2} & 1 - \dfrac{X}{1 - Y^2} & -i\sin\theta\dfrac{XY}{1 - Y^2} \\ -\sin\theta\cos\theta\dfrac{XY^2}{1 - Y^2} & i\sin\theta\dfrac{XY}{1 - Y^2} & 1 - \dfrac{X(1 - Y^2 \cos^2\theta)}{1 - Y^2} \end{pmatrix}, \tag{2.54}$$

The three components of the wave equation (2.31) are

$$-k^2 E_1' + \frac{\omega^2}{c^2}(\varepsilon_{11}' E_1' + \varepsilon_{12}' E_2' + \varepsilon_{13}' E_3') = 0,$$

$$-k^2 E_2' + \frac{\omega^2}{c^2}(\varepsilon_{21}' E_1' + \varepsilon_{22}' E_2' + \varepsilon_{23}' E_3') = 0, \qquad (2.55)$$

$$\frac{\omega^2}{c^2}(\varepsilon_{31}' E_1' + \varepsilon_{32}' E_2' + \varepsilon_{33}' E_3') = 0.$$

By eliminating k from the first two equations and replacing E_3' from the third, we obtain the equation for the wave polarization $\rho \equiv E_1'/E_2'$

$$\rho^2(\varepsilon_{21}'\varepsilon_{33}' - \varepsilon_{23}'\varepsilon_{31}') + \rho(\varepsilon_{22}'\varepsilon_{33}' - \varepsilon_{23}'\varepsilon_{32}' - \varepsilon_{11}'\varepsilon_{33}' + \varepsilon_{13}'\varepsilon_{31}')$$

$$- \varepsilon_{12}'\varepsilon_{33}' + \varepsilon_{13}'\varepsilon_{32}' = 0, \qquad (2.56)$$

which by using (2.54) can be written as

$$\rho^2 - \frac{iY\sin^2\theta}{(1-X)\cos\theta}\rho + 1 = 0. \qquad (2.57)$$

The two solutions are

$$\rho_\pm = i\frac{Y\sin^2\theta}{2(1-X)\cos\theta} \pm i\left[1 + \frac{Y^2\sin^4\theta}{4(1-X)^2\cos^2\theta}\right]^{1/2}, \qquad (2.58)$$

with

$$\rho_+\rho_- = 1, \qquad (2.59)$$

and

$$|\rho_+| > |\rho_-|. \qquad (2.60)$$

On the (x_1', x_2') plane, then, the real components Re of the electric field can be written as

$$\text{Re } E_1' = |\rho_\pm|\cos(\mathbf{k}\cdot\mathbf{r} - \omega t \pm \pi/2),$$

$$\text{Re } E_2' = \cos(\mathbf{k}\cdot\mathbf{r} - \omega t), \qquad (2.61)$$

showing that the wave is elliptically polarized with the ellipse having the major axis parallel to either the x_1' axis (ρ_+) or the x_2 axis (ρ_-), so that the two modes have orthogonal polarizations. For $\theta \to \pi/2$,

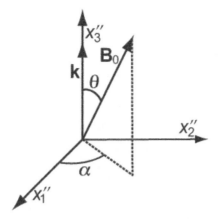

Figure 2.5. System of coordinates with arbitrary direction of the magnetic field.

i.e., when $\mathbf{B}_0$ is parallel to the x_1' axis, $\rho_+ \rightarrow \infty(E_2' \rightarrow 0)$ and $\rho_- \rightarrow 0$ $(E_1' \rightarrow 0)$. When $\theta = 0$, i.e., when $\mathbf{B}_0$ is parallel to the x_3' axis, $\rho_\pm = \pm i$ (circular polarization). This agrees with the previous assessment of wave polarization in Sec. 2.3 if the $\pm$subscript in (2.58) has the same meaning as in (2.43).

Finally, in the new system of coordinates (x_1'', x_2'', x_3'') formed by a rotation of the coordinates of Fig. 2.2 around the x_3' axis by the angle $-\alpha$ (Fig. 2.5), we have

$$E_1'' = E_1' \cos \alpha - E_2' \sin \alpha,$$
$$E_2'' = E_1' \sin \alpha + E_2' \cos \alpha, \tag{2.62}$$

from which, using $E_1' = \rho_\pm E_2'$, we obtain

$$\frac{E_1''}{E_2''} = \frac{\rho_\pm - \tan \alpha}{1 + \rho_\pm \tan \alpha}, \tag{2.63}$$

giving the polarization for an arbitrary direction of the magnetic field.

2.5 Wave Packets in Homogeneous Plasmas

So far we have considered only plane waves with a well-defined frequency and wave vector. In general, however, we must deal with more complicated wave-like phenomena, which in homogeneous plasmas can be represented by integrals of the form

$$\mathbf{E}(\mathbf{r}, t) = \int \mathbf{e}(\mathbf{k}) e^{i\psi(\mathbf{r},t,\mathbf{k})} d\mathbf{k},$$

$$\psi(\mathbf{r}, t, \mathbf{k}) = \mathbf{k} \cdot \mathbf{r} - \Omega(\mathbf{k})t, \tag{2.64}$$

where $\omega = \Omega(\mathbf{k})$ is derived from the dispersion relation (2.41). The wave packet may comprise two integrals of this form, one for each mode of propagation. In general, integrals of this type cannot be evaluated in closed form and one must resort to approximation techniques.

For large values of $\mathbf{r}$ and t, the major contribution to the integral arises from the vicinity of the points $\mathbf{k}_s$ of stationary phase defined by $\nabla_k \psi = 0$ (where ∇_k is the gradient operator in $\mathbf{k}$), or

$$\mathbf{v}_G(\mathbf{k}) \equiv \nabla_k \Omega(\mathbf{k}) = \frac{\mathbf{r}}{t} \quad \text{at } \mathbf{k} = \mathbf{k}_s(\mathbf{r}, t). \tag{2.65}$$

Near $\mathbf{k}_s$, the slowly varying amplitude may be approximated by $\mathbf{e}(\mathbf{k}) \approx \mathbf{e}(\mathbf{k}_s)$, while the phase requires a power series expansion of up to quadratic terms in $(\mathbf{k} - \mathbf{k}_s)$, resulting in [6]

$$\mathbf{E}(\mathbf{r}, t) \approx \mathbf{e}(\mathbf{k}_s) \left(\frac{2\pi}{t}\right)^{3/2} G^{-1/2} \exp[i(\mathbf{k} \cdot \mathbf{r} - \Omega(\mathbf{k})t + \varphi)], \tag{2.66}$$

where φ is a phase that depends on the structure of Ω near the points of stationary phase and G is the determinant of the matrix $|\nabla_k \nabla_k \Omega(\mathbf{k})|$ at $\mathbf{k} = \mathbf{k}_s$. The phase ψ is

$$\psi(\mathbf{r}, t) = \mathbf{k}(\mathbf{r}, t) \cdot \mathbf{r} - \Omega(\mathbf{k})t, \tag{2.67}$$

from which we get

$$\nabla \psi = \mathbf{k} + [\mathbf{r} - t\nabla_k \Omega(\mathbf{k})] \cdot \nabla \mathbf{k},$$

$$\frac{\partial \psi}{\partial t} = -\Omega(\mathbf{k}) + [\mathbf{r} - t\nabla_k \Omega(\mathbf{k})] \frac{\partial \mathbf{k}}{\partial t}. \tag{2.68}$$

In a frame of reference moving at constant velocity $d\mathbf{r}/dt = \nabla_k \Omega$, these becomes

$$\nabla \psi = \mathbf{k}_s,$$

$$\frac{\partial \psi}{\partial t} = -\Omega(\mathbf{k}_s). \tag{2.69}$$

From this, we conclude that an observer moving with the wave packet at the constant velocity (the *group velocity*)

$$\mathbf{v}_G = \nabla_k \Omega, \qquad (2.70)$$

would see a constant wave number $\mathbf{k}_s$ and constant frequency $\Omega(\mathbf{k}_s)$. Or in other words, in homogeneous plasmas, a trajectory on which the frequency ω of a wave packet remains constant is defined in $(\mathbf{r}, \mathbf{k})$ space by the equations

$$\frac{d\mathbf{r}}{dt} = \mathbf{v}_G,$$
$$\frac{d\mathbf{k}}{dt} = 0. \qquad (2.71)$$

These may be considered the ray equations in homogeneous plasmas.

References

[1] Klimontovich, Yu. L., *The Statistical Theory of Non-equilibrium Processes in a Plasma*, M.I.T. Press, Cambridge, 1967.

[2] De Groot, S. R. and Mazur, P., *Non-equilibrium Thermodynamics*, Dover, New York, 1985.

[3] Stix, T. H., *Waves in Plasmas*, American Institute of Physics, New York, 1992.

[4] Budden K. G., *Radio Waves in the Ionosphere*, Cambridge University Press, Cambridge, 1962.

[5] Ginzburg, V. L., *The Propagation of Electromagnetic Waves in Plasmas*, Pergamon Press, Oxford, 1965.

[6] Felsen, L. B. and Marcuwitz, N., *Radiation and Scattering of Waves*, Prentice-Hall, Inc., Englewood Cliffs, NJ, 1974.

CHAPTER 3

INHOMOGENEOUS PLASMAS

In laboratory experiments, one deals with confined plasmas that — by definition — are not homogeneous media. In this chapter, we first review the standard approximation of geometrical optics for wave propagation in inhomogeneous plasmas, where the wave is viewed as composed of independent rays, each propagating along a trajectory where the local dispersion relation is satisfied everywhere.

This approximation provides a simple and elegant technique for accounting refractive effects in non-homogeneous plasmas, and plays an important role in the interpretation of measurements using electromagnetic waves. However, just because of the independence of rays, this approximation may lead to non-physical results, such as the focusing of the wave on a single point. This is obviously due to the fact that diffraction is not included in the geometrical approximation. In the second part of this chapter, we discuss a refinement of the theory of geometrical optics including lowest-order diffractive effects.

3.1 Wave Packets in Weakly Inhomogeneous Plasmas

When the properties of the medium vary in space and time, an expansion of the field in terms of plane waves is no longer convenient since the latter do not represent the eigenfunctions of the system. However, the case of interest in plasma diagnostics is that of weakly inhomogeneous media, where the relative change of plasma parameters over an electromagnetic scale length is small. In this case, the asymptotic form of the phase function (2.67) in homogeneous plasmas suggests a solution of Maxwell equations

$$\nabla \times \mathbf{B} = \frac{1}{c}\frac{\partial \mathbf{E}}{\partial t} + \frac{4\pi}{c}\mathbf{j},$$

$$\nabla \times \mathbf{E} = -\frac{1}{c}\frac{\partial \mathbf{B}}{\partial t}, \tag{3.1}$$

in the form [1, 2]

$$\mathbf{E}(\mathbf{r}, t) = \mathbf{e}\,e^{i\psi(\mathbf{r},t)},$$

$$\mathbf{B}(\mathbf{r}, t) = \mathbf{b}\,e^{i\psi(\mathbf{r},t)}, \tag{3.2}$$

where $\mathbf{e}$ and $\mathbf{b}$ are slowly varying functions, and ψ — the *eikonal* — plays the role of the wave phase. By defining

$$\mathbf{k}(\mathbf{r}, t) = \nabla\psi,$$

$$\omega(\mathbf{r}, t) = -\frac{\partial\psi}{\partial t}, \tag{3.3}$$

as the local wave vector and frequency, (3.2) acquires a form which is reminiscent of similar equations in uniform plasmas. Equations (3.1) then become

$$\nabla \times \mathbf{b} + i\mathbf{k} \times \mathbf{b} = \frac{1}{c}\left(\frac{\partial \mathbf{e}}{\partial t} - i\omega\mathbf{e}\right) + \frac{4\pi}{c}\boldsymbol{\sigma} \cdot \mathbf{e},$$

$$\nabla \times \mathbf{e} + i\mathbf{k} \times \mathbf{e} = -\frac{1}{c}\left(\frac{\partial \mathbf{b}}{\partial t} - i\omega\mathbf{b}\right), \tag{3.4}$$

where $\boldsymbol{\sigma}$ is the conductivity tensor.

To seek an asymptotic solution of (3.1), we formally expand the field amplitudes $\mathbf{e}$ and $\mathbf{b}$ in ascending powers of a small dimensionless parameter δ (a measure of the fractional change of the background plasma over a wavelength or a wave period)

$$\mathbf{e} = \mathbf{e}_0 + \mathbf{e}_1 + \cdots \quad \mathbf{b} = \mathbf{b}_0 + \mathbf{b}_1 + \cdots. \tag{3.5}$$

When these are substituted in (3.4) and coefficients of like powers are equated, the zero-order yields

$$ck \times b_0 + \omega \boldsymbol{\varepsilon}_0 \cdot e_0 = 0,$$
$$ck \times e_0 - \omega b_0 = 0, \tag{3.6}$$

where $\boldsymbol{\varepsilon}_0 = \mathbf{I} + (4\pi i/\omega)\boldsymbol{\sigma}_0$, with $\boldsymbol{\sigma}_0$ the lowest-order anti-Hermitian plasma conductivity. Similarly, the next order equations are obtained by equating to zero the first-order terms in (3.4)

$$i\left(\mathbf{k} \times \mathbf{b}_1 + \frac{\omega}{c}\boldsymbol{\varepsilon}_0 \cdot \mathbf{e}_1\right) = -\frac{4\pi i}{c}\boldsymbol{\sigma}_1 \cdot \mathbf{e}_0 - \nabla \times \mathbf{b}_0 + \frac{1}{c}\frac{\partial \mathbf{e}_0}{\partial t},$$
$$i\left(\mathbf{k} \times \mathbf{e}_1 - \frac{\omega}{c}\mathbf{b}_1\right) = -\nabla \times \mathbf{e}_0 - \frac{1}{c}\frac{\partial \mathbf{b}_0}{\partial t}, \tag{3.7}$$

where $\boldsymbol{\sigma}_1$ is the first-order plasma conductivity, including a Hermitian component that is responsible for plasma absorption, and terms due to the slowly varying medium. Calculation of the latter requires lengthy vector manipulations [2, 3] that are beyond the scope of this chapter.

Using the second of equations (3.6) to eliminate $\mathbf{b}_0$ from the first yields

$$\mathbf{k} \times (\mathbf{k} \times \mathbf{e}_0) + \frac{\omega^2}{c^2}\boldsymbol{\varepsilon}_0 \cdot \mathbf{e}_0 = 0. \tag{3.8}$$

Similarly to the case of homogeneous plasmas, the condition for existence of non-trivial solutions of this equation results in a local dispersion relation

$$\omega = \Omega(\mathbf{k}, \mathbf{r}, t), \tag{3.9}$$

to be satisfied with $\mathbf{k}$ and ω given by (3.3).

We can quickly derive the ray equation in weakly inhomogeneous plasmas by noting that the cross-differentiation of (3.3) yields

$$\frac{\partial \omega}{\partial x_i} + \frac{\partial k_i}{\partial t} = 0, \tag{3.10}$$

$$\frac{\partial k_i}{\partial x_j} - \frac{\partial k_j}{\partial x_i} = 0, \tag{3.11}$$

in the system of orthogonal coordinates with $\mathbf{r} = (x_1, x_2, x_3)$ and $\mathbf{k} = (k_1, k_2, k_3)$. Substituting (3.9) in (3.10) gives

$$\frac{\partial \Omega}{\partial x_i} + \frac{\partial \Omega}{\partial k_j}\frac{\partial k_j}{\partial x_i} + \frac{\partial k_i}{\partial t} = 0, \tag{3.12}$$

with the usual convention on repeated indexes. From this and (3.11) we obtain (using the notation ∇_r and ∇_k for the gradient operators in $\mathbf{r}$ and $\mathbf{k}$)

$$\frac{\partial \mathbf{k}}{\partial t} + \mathbf{v}_G \cdot \nabla \mathbf{k} = -\nabla_r \Omega, \tag{3.13}$$

where $\mathbf{v}_G$ is the group velocity as defined by

$$\mathbf{v}_G(\mathbf{k}, \mathbf{r}, t) = \nabla_k \Omega(\mathbf{k}, \mathbf{r}, t), \tag{3.14}$$

as in (2.70). Equation (3.13) may be viewed as giving the evolution of $\mathbf{k}$ as seen by an observer moving at the local group velocity, and thus the wave ray equations in a weakly inhomogeneous plasma can be cast in canonical form

$$\frac{d\mathbf{r}}{dt} = \nabla_k \Omega,$$

$$\frac{d\mathbf{k}}{dt} = -\nabla_r \Omega, \tag{3.15}$$

with Ω playing the role of the Hamiltonian. These equations together with (3.9) give

$$\frac{d\omega}{dt} = \frac{\partial \Omega}{\partial t}. \tag{3.16}$$

Finally, the phase difference between two points s_1 and s_2 of the same ray is

$$\Delta\psi = \int_{s_1}^{s_2} \mathbf{k} \cdot d\mathbf{s}, \tag{3.17}$$

where $d\mathbf{s}$ is the arc element along the ray. These are the equations of geometrical optics.

The derivatives of Ω in the ray equations can be obtained from differentiation of the dispersion relation $D(\omega, \mathbf{k}, \mathbf{r}, t) = 0$ with respect to **r** and **k**

$$\nabla_r D + \frac{\partial D}{\partial \omega} \nabla_r \Omega = 0,$$

$$\nabla_k D + \frac{\partial D}{\partial \omega} \nabla_k \Omega = 0,$$

(3.18)

which transforms the ray equations into

$$\frac{d\mathbf{r}}{dt} = -\frac{\nabla_k D}{\partial D / \partial \omega},$$

$$\frac{d\mathbf{k}}{dt} = \frac{\nabla_r D}{\partial D / \partial \omega}.$$

(3.19)

3.2 Ray Equations in Absorbing Plasmas

The aim of this section is to show that the approximation of geometrical optics extends beyond the case of cold plasmas where **k** is real and the wave absorption is negligible. This is a critical issue since the ray equations (3.15) find its most important use in the diagnostic of hot plasmas where kinetic effects produce an anti-Hermitian component of the dielectric tensor [4], and the solution of the dispersion relation becomes a complex wave vector $\mathbf{k} = \mathbf{k}_r + i\mathbf{k}_i$ (where $|\mathbf{k}_r| \gg |\mathbf{k}_i|$ is usually the only case of interest for plasma diagnostics). Under these conditions, it is possible to reformulate the theory of geometrical optics with new ray equations that are expressed in terms of the real part of the dispersion relation and $\mathbf{k}_r$ [2, 5, 6].

At first sight, the procedure that we have used in the previous section is not applicable to hot plasmas since in general the imaginary part of D is not much smaller than the real part even in the case of weak absorption, and thus the dispersion relation cannot lead directly to an equation for $\mathbf{k}_r$. However, similarly to (2.41), the dispersion relation continues to be a polynomial of fourth degree in $N_\perp$

$$D = \varepsilon_{11} N_\perp^4 + P = 0, \qquad (3.20)$$

where D (including ε_{11}) is now complex and P is a polynomial of third degree in $N_\perp$ [7, 8]. Assuming $N_\perp$ to be always finite and

$\varepsilon_{11} \neq 0$, (3.20) is equivalent to

$$F = D/\varepsilon_{11} = N_\perp^4 + P/\varepsilon_{11} = 0, \qquad (3.21)$$

and thus, from the assumption that $|\mathbf{k}_r| \gg |\mathbf{k}_i|$, we get that the real part of P/ε_{11} is much larger than its imaginary part, so that we can write

$$F(\omega, \mathbf{k}, \mathbf{r}, t) \approx F_r(\omega, \mathbf{k}_r, \mathbf{r}, t) + iF_i(\omega, \mathbf{k}_r, \mathbf{r}, t) + i\mathbf{k}_i \cdot \nabla_{k_r} F_r = 0,$$
$$(3.22)$$

from which

$$F_r(\omega, \mathbf{k}_r, \mathbf{r}, t) = 0,$$
$$\mathbf{k}_i \cdot \nabla_{k_r} F_r = -F_i(\omega, \mathbf{k}_r, \mathbf{r}, t). \qquad (3.23)$$

The first of these equations yields ray equations similar to (3.19)

$$\frac{d\mathbf{r}}{dt} = \mathbf{v}_{\mathrm{G}} = -\frac{\nabla_{k_r} F_r}{\partial F_r/\partial \omega},$$
$$\frac{d\mathbf{k}_r}{dt} = \frac{\nabla_r F_r}{\partial F_r/\partial \omega}, \qquad (3.24)$$

while the second gives

$$\int_{s_1}^{s_2} \mathbf{k}_i \cdot ds = \int_{s_1}^{s_2} (\mathbf{k}_i \cdot \mathbf{v}_{\mathrm{G}}/v_G) ds = \int_{s_1}^{s_2} \frac{|F_i|}{|\nabla_{k_r} F_r|} ds, \qquad (3.25)$$

which can be used for calculating the wave absorption between two points of a ray [9].

3.3   Ray Tracing

As already mentioned, the approximation of geometrical optics plays an important role in the interpretation of measurements using electromagnetic waves. In this section, the ray equations (3.19) are recast in a form suitable for numerical calculations, and a few examples of ray tracing are shown.

We consider a time-independent plasma in a magnetic field $\mathbf{B}$, with the Appleton–Hartree dispersion relation

$$D(\omega, \mathbf{k}, \mathbf{r}) = 1 - \frac{k^2 c^2}{\omega^2} - \frac{2X(1-X)}{2(1-X) - Y^2 \sin^2 \theta \pm \Gamma} = 0, \qquad (3.26)$$

where $\Gamma = [Y^4 \sin^4 \theta + 4Y^2(1-X)^2 \cos^2 \theta]^{1/2}$, and the $\pm$ sign corresponds to the ordinary $(+)$ and the extraordinary $(-)$ mode. For such a medium, it is convenient to use the arc length s along the ray as a parameter and express the ray equations (3.19) in the form

$$\frac{d\mathbf{r}}{ds} = \frac{1}{v_G}\frac{d\mathbf{r}}{dt} = -\mathrm{sgn}\left(\frac{\partial D}{\partial \omega}\right)\frac{\nabla_k D}{|\nabla_k D|},$$

$$\frac{d\mathbf{k}}{ds} = \frac{1}{v_G}\frac{d\mathbf{k}}{dt} = \mathrm{sgn}\left(\frac{\partial D}{\partial \omega}\right)\frac{\nabla_r D}{|\nabla_k D|}.$$

(3.27)

Following [10], we resolve the group velocity into components along $\mathbf{e}_k = \mathbf{k}/k$ and $\mathbf{e}_\theta = \mathbf{k}\times(\mathbf{k}\times\mathbf{B})/|\mathbf{k}\times\mathbf{k}\times\mathbf{B})|$, casting the ray equations into the form

$$\frac{d\mathbf{r}}{ds} = -\mathrm{sgn}\left(\frac{\partial D}{\partial \omega}\right)\frac{\frac{\partial D}{\partial k}\mathbf{e}_k + \frac{1}{k}\frac{\partial D}{\partial \theta}\mathbf{e}_\theta}{\left[\left(\frac{\partial D}{\partial k}\right)^2 + \left(\frac{1}{k}\frac{\partial D}{\partial \theta}\right)^2\right]^{1/2}},$$

$$\frac{d\mathbf{k}}{ds} = \mathrm{sgn}\left(\frac{\partial D}{\partial \omega}\right)\frac{\frac{\partial D}{\partial X}\nabla X + \frac{\partial D}{\partial Y^2}\nabla Y^2 + \frac{\partial D}{\partial \theta}\nabla\theta}{\left[\left(\frac{\partial D}{\partial k}\right)^2 + \left(\frac{1}{k}\frac{\partial D}{\partial \theta}\right)^2\right]^{1/2}}.$$

(3.28)

For the sake of easier notations, we define the functions

$$A(X) = 2X(1-X),$$
$$B_m(X, Y^2, \theta) = 2(1-X) - Y^2\sin^2\theta + m\Gamma,$$
$$G = A/B_m,$$

(3.29)

where $m = \pm 1$, and thus

$$D(k, \theta, X, Y^2, \omega) = 1 - \frac{k^2 c^2}{\omega^2} - G.$$

(3.30)

All derivatives on the right-hand side of (3.28) are

$$\frac{\partial D}{\partial k} = -\frac{2kc^2}{\omega^2},$$

$$\frac{\partial D}{\partial \theta} = \frac{A}{B_m^2}\frac{\partial B_m}{\partial \theta} = -G\frac{2\sin\theta\cos\theta}{B_m}Y^2$$

$$\times\left[1 - m\frac{Y^2\sin^2\theta - 2(1-X)^2}{\Gamma}\right],$$

$$\frac{\partial D}{\partial X} = -\frac{\partial G}{\partial X} - \frac{\partial G}{\partial \Gamma}\frac{\partial \Gamma}{\partial X}, \quad \frac{\partial D}{\partial Y^2} = -\frac{\partial G}{\partial Y^2} - \frac{\partial G}{\partial \Gamma}\frac{\partial \Gamma}{\partial Y^2},$$

$$\frac{\partial G}{\partial X} + \frac{\partial G}{\partial \Gamma}\frac{\partial \Gamma}{\partial X} = \frac{1}{B_m^2}\left[2(1-2X)B_m + 2A + mA\frac{4Y^2(1-X)}{\Gamma}\cos^2\theta\right],$$

$$\frac{\partial G}{\partial Y^2} + \frac{\partial G}{\partial \Gamma}\frac{\partial \Gamma}{\partial Y^2} = \frac{A}{B_m^2}\left[\sin^2\theta - m\frac{Y^2\sin^4 + 2(1-X)^2\cos^2\theta}{\Gamma}\right],$$

$$\frac{\partial D}{\partial \omega} = \frac{2k^2c^2}{\omega^3} + \frac{2X}{\omega}\left(\frac{\partial G}{\partial X} + \frac{\partial G}{\partial \Gamma}\frac{\partial \Gamma}{\partial X}\right)$$

$$+ \frac{2Y^2}{\omega}\left(\frac{\partial G}{\partial Y^2} + \frac{\partial G}{\partial \Gamma}\frac{\partial \Gamma}{\partial Y^2}\right),$$

$$\nabla X = X\frac{\nabla n_e}{n_e}, \quad \nabla Y^2 = 2Y^2\frac{\nabla B}{B},$$

$$\nabla\theta = -\frac{1}{k\sin\theta}\nabla\left(\mathbf{k}\cdot\frac{\mathbf{B}}{B}\right),$$

where in the last equation the gradient acts only on the magnetic field.

As an example of ray tracing we use the tokamak configuration of Fig. 1.6(a) with the density profile and characteristic frequencies of Fig. 3.1. Then, the ray equations (3.28) — a system of ordinary differential equations — can be solved numerically with the Runge–Kutta method [11]. The results are in Fig. 3.2 showing the ray trajectories of waves with a frequency of 165 GHz that are launched from two different locations with either the *ordinary* (solid lines) or the *extraordinary* (dash lines) mode of propagation, and in both cases with the initial condition $k_\phi = 0$. For the extraordinary mode, the rays are reflected (left frame) or strongly bended (right frame) because of the X-mode upper cutoff.

3.4   The Complex Eikonal Approximation

In order to prevent the collapse of rays into a focal point or a caustic, Choudhary and Felsen [12] proposed to modify the formulation of geometrical optics by using a complex *eikonal*. To see how drastically this can modify the ray equations, let us begin by taking the wave

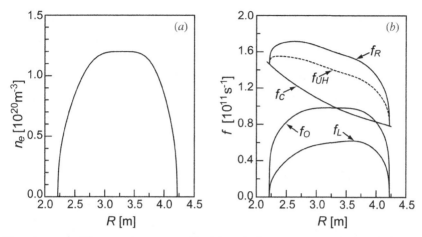

Figure 3.1. Electron density profile (a) and characteristic frequencies (b) on the equatorial plane of the tokamak of Fig. 1.6(a) with $B_T = 3.5\,\mathrm{T}$ ($f_c = \omega_c/2\pi$, $f_O = \omega_O/2\pi$, $f_R = \omega_R/2\pi$, $f_L = \omega_L/2\pi$, $f_{UH} = \omega_{UH}/2\pi$).

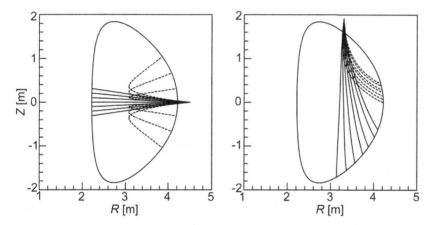

Figure 3.2. Ray trajectories of waves with frequency of 165 GHz launched from two different locations; solid lines are for the O-mode and dash lines for the X-mode.

electric field in the form

$$\mathbf{E}(\mathbf{r}, t) = \mathbf{E}_0(\mathbf{r}) \exp(ik_0 S(\mathbf{r}) - i\omega t), \qquad (3.31)$$

where ω is a real frequency, $k_0 = \omega/c$ is the wave number in free-space and the function $S(\mathbf{r}) = S_r(\mathbf{r}) + iS_i(\mathbf{r})$ is the complex *eikonal*.

Formally, we can define $k_0 \nabla S$ as a complex wave vector

$$\mathbf{k} = \mathbf{k}_r + i\mathbf{k}_i, \tag{3.32}$$

where in the following we will assume $k_r \gg k_i$.

As in the theory of geometrical optics, $\mathbf{E}_0(\mathbf{r})$ is assumed to have a slowly varying amplitude that can be expressed as a series of powers in $1/k_0$

$$\mathbf{E}_0(\mathbf{r}) = \sum_{n \geq 0} \frac{\mathbf{e}_n(\mathbf{r})}{(ik_0)^n}, \tag{3.33}$$

where $\mathbf{e}_n$ are complex functions of position. Following the same procedure that led to (3.9), from the zero order terms we obtain the local dispersion relation for the case of complex *eikonal*

$$D(\mathbf{r}, \nabla S, \omega) \equiv (\nabla S)^2 - N^2(\mathbf{r}, \nabla S, \omega) = 0, \tag{3.34}$$

where N plays the role of a complex refractive index. By expanding N^2 around $\mathbf{k}_r$ and keeping terms of the first order in k_r/k_i, we obtain [13]

$$D(\mathbf{r}, \mathbf{k}_r, \omega) + i\mathbf{k}_i \cdot \frac{\partial D(\mathbf{r}, \mathbf{k}_r, \omega)}{\partial \mathbf{k}_r} - \frac{1}{2}\mathbf{k}_i\mathbf{k}_i : \frac{\partial^2 D(\mathbf{r}, \mathbf{k}_r, \omega)}{\partial \mathbf{k}_r \partial \mathbf{k}_r} = 0, \tag{3.35}$$

where use is made of the dyadic notation. For a loss-free plasma (i.e., when N^2 is a real function), separation of (3.35) into real and imaginary parts gives

$$H(\mathbf{r}, \mathbf{k}_r, \omega) \equiv N^2(\mathbf{r}, \mathbf{k}_r, \omega) - \frac{c^2}{\omega^2}(k_r^2 - k_i^2)$$

$$-\frac{1}{2}\mathbf{k}_i\mathbf{k}_i : \frac{\partial^2 N^2(\mathbf{r}, \mathbf{k}_r, \omega)}{\partial \mathbf{k}_r \partial \mathbf{k}_r} = 0, \tag{3.36}$$

$$\mathbf{k}_i \cdot \left(2\frac{c^2}{\omega^2}\mathbf{k}_r - \frac{\partial N^2(\mathbf{r}, \mathbf{k}_r, \omega)}{\partial \mathbf{k}_r} \right) = 0.$$

The first of these equations becomes much simpler when, as in the case of the Appleton–Hartree formula, the refractive index has the form $N = N(\mathbf{r}, N_{||}, \omega)$ with $N_{||} = (\mathbf{k}_r/k_0) \cdot \mathbf{b}$ and $\mathbf{b} = \mathbf{B}/B$, so that

$$H(\mathbf{r}, \mathbf{k}_r, \omega) = N^2(\mathbf{r}, N_{||}, \omega) - \frac{c^2}{\omega^2} k_r^2 + (\nabla S_i)^2$$

$$- \frac{1}{2} (\nabla S_i \cdot \mathbf{b})^2 \frac{\partial^2 N^2(\mathbf{r}, N_{||}, \omega)}{\partial^2 N_{||}}. \qquad (3.37)$$

The first of (3.36) leads to the ray equations [13],

$$\frac{d\mathbf{r}}{dt} = -\frac{\nabla_{k_r} H}{\partial H / \partial \omega} = \mathbf{v}_G,$$

$$\frac{d\mathbf{k}_r}{dt} = \frac{\nabla_r H}{\partial H / \partial \omega}. \qquad (3.38)$$

These must be solved self-consistently with the second of (3.36), which can be cast in the form

$$\nabla S_i \cdot \frac{d\mathbf{r}}{dt} = 0, \qquad (3.39)$$

indicating that the imaginary component of the *eikonal* is constant along the ray trajectories. The presence of $(\nabla S_i)^2$ in these equations is the essence of the complex *eikonal* approximation.

Finally, as in (3.27), it is convenient to use the arc length s along the ray as a parameter in the ray equations (3.38), which become

$$\frac{d\mathbf{r}}{ds} = \frac{1}{v_G} \frac{d\mathbf{r}}{dt} = -\mathrm{sgn}\left(\frac{\partial H}{\partial \omega}\right) \frac{\nabla_{k_r} H}{|\nabla_{k_r} H|},$$

$$\frac{d\mathbf{k}_r}{ds} = \frac{1}{v_G} \frac{d\mathbf{k}_r}{dt} = \mathrm{sgn}\left(\frac{\partial H}{\partial \omega}\right) \frac{\nabla_r H}{|\nabla_{k_r} H|}. \qquad (3.40)$$

3.5 Propagation of a Gaussian Beam

In this section, a few examples of numerical solutions of ray equations (3.40) are described for the case of a circular cross-section Gaussian beam, launched into the plasma from free space.

Equations (3.40) are first-order ordinary differential equations, which are coupled by terms in ∇S_i in the dispersion relation (3.35). This is what prevents the crossing of rays and the formation of focal points. These equations must be solved together with (3.39), imposing the constancy of S_i along the ray trajectories.

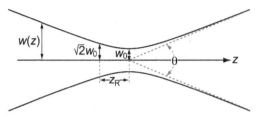

Figure 3.3.   Width of a Gaussian beam $w(z)$ as a function of axial distance z; w_0 is the beam waist, z_R is the Rayleigh length and θ is the total beam divergence.

The numerical scheme for the integration of (3.40) can be briefly described as follows [13–15]. The Gaussian beam is first split into a set of many rays, whose initial conditions can be obtained from the standard paraxial approximation of a Gaussian beam [16], giving

$$E(x,y,z) \propto \exp\left(-\frac{x^2+y^2}{w(z)^2}\right)\exp\left(ik_0\left(z+\frac{x^2+y^2}{2R(z)}\right)\right), \quad (3.41)$$

where (x,y,z) is a system of orthogonal coordinates with the z-axis along the direction of propagation (Fig. 3.3). In (3.41), $w(z) = w_0[1+(z/z_R)^2]^{1/2}$ is the beam radius and $R(z) = (z^2+z_R^2)/z$ is the radius of curvature of the wave front, with w_0 the beam waist and $z_R = k_0 w_0^2/2$ the Rayleigh length. For $z \gg z_R$, w increases linearly with z, and thus the beam diverge is $\theta \approx 4/k_0 w_0$.

From (3.41), we get the initial conditions

$$S_r = z + \frac{x^2+y^2}{2R}, \quad S_i = \frac{x^2+y^2}{k_0 w^2}, \quad (3.42)$$

from which we get those for $\mathbf{k}_r$ and ∇S_i.

Because of the second of (3.42), ∇S_i is zero on the central ray. This decouples its trajectory from that of the other rays, which there-fore can be obtained from the equations of geometrical optics (3.27). As for the other rays, starting from the launching plane where ∇S_i and its derivatives can be obtained analytically from (3.42), they are advanced by a given integration step and thus the initial values of S_i are transported to a new cluster of points. The first and second derivatives of S_i in a system of orthogonal coordinates (x_1, x_2, x_3) can then be obtained with the following procedure. Among the cluster of points, let's choose two nearby points P_1 and P_2, and for a generic function of position $f(\mathbf{x})$ define $\Delta f = f(P_2) - f(P_1)$. At the lowest order in $|\Delta\mathbf{x}|$, we get

$$\Delta S_i = \frac{\partial S_i}{\partial x_1}\Delta x_1 + \frac{\partial S_i}{\partial x_2}\Delta x_2 + \frac{\partial S_i}{\partial x_3}\Delta x_3. \tag{3.43}$$

By repeating the process using three different points P_2, one obtains a system of three linear equations in the unknown first partial derivatives of S_i, whose solution gives the value of ∇S_i at P_1. The same procedure using the known values of ∇S_i provides the second partial derivatives of S_i. This allows advancing the rays to the next step, and thus to get the full solution of (3.40) by iteration.

Figure 3.4, which displays the propagation in free-space of a Gaussian beam with a frequency of 75 GHz and an initial radius $w = 7.5$ cm, shows that such a numerical procedure yields the complete solution of (3.40). Here we can see the typical behavior of a Gaussian beam where its radius decreases to a minimum — the beam waist w_0 — and then begins to diverge. As mentioned above, when the distance L from the waist is larger than the Rayleigh length,

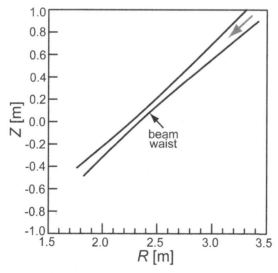

Figure 3.4. Ray tracing from the complex eikonal approximation (3.40) for a Gaussian beam in free space. The beam is launched parallel to the $(r,\,z)$-plane with a frequency of 75 GHz and an initial radius $w = 7.5$ cm. The rays shown are those with an e-folding amplitude.

as in this case, the divergence of a Gaussian beam in free space is $\theta = 4/k_0 w_0$, so that $w = 2L/k_0 w_0$. For the case of Fig. 3.4, where $w = 7.5\,\text{cm}$ at $130\,\text{cm}$ from the waist, we obtain $w_0 = 2.2\,\text{cm}$ that is in good agreement with the numerical result.

Figure 3.5 shows the ray tracing from the complex eikonal approximation (3.40) for the same Gaussian beam of Fig. 3.4 in a circular cross-section tokamak plasma with a toroidal magnetic field of $5.2\,\text{T}$ and a central electron density of $4 \times 10^{19}\,\text{m}^{-3}$. The comparison

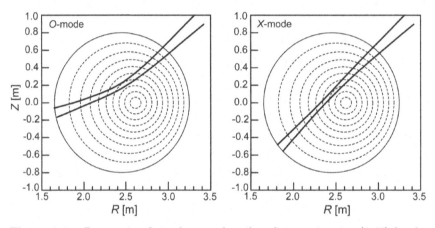

Figure 3.5. Ray tracing from the complex eikonal approximation (3.40) for the same beam of Fig. 3.4 in a circular cross-section tokamak plasma (dash lines are magnetic surfaces).

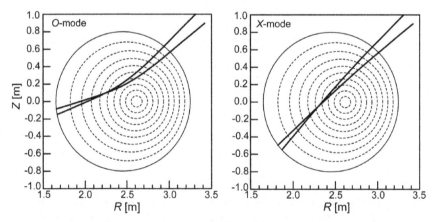

Figure 3.6. Same as in Fig. 3.5 from the standard geometrical optics approximation (3.27).

of these results with those from standard geometrical optics (Fig. 3.6) demonstrates how the complex eikonal approximation prevents the formation of focal points.

In conclusion, extension of the geometrical optics approximation to the case of complex eikonals provides a simple and powerful tool for taking into account lowest-order diffractive effects in complex inhomogeneous plasmas.

Bibliography

[1] Weinberg, S., *Phys. Rev.* **126**, 1899 (1962).
[2] Bernstein, I. B., *Phys. Fluids* **18**, 320 (1975).
[3] Bravo-Ortega, A. and Glasser, A. H., *Phys. Fluids B* **3**, 529 (1991).
[4] Stix, T. H., *Waves in Plasmas*, American Institute of Physics, New York, 1993.
[5] Suchy, K. J., *Plasma Phys.* **8**, 53 (1972).
[6] Fidone I., Granata, G. and Meyer, R. L., *Phys. Fluids* **25**, 2249 (1982).
[7] Bornatici, M., Cano, R., De Barbieri, O. and Engelmann, F., *Nucl. Fusion* **23**, 1153 (1983).
[8] Fidone I., Giruzzi G., Krivenski, V. and Ziebell, L. F., *Nucl. Fusion* **26**, 1537 (1986).
[9] Mazzucato, E., Fidone, I. and Granata, G., *Phys. Fluids* **30**, 3745 (1987).
[10] Batchelor, D. B. and Goldfinger, R. C., ORNL/TM-6844 (1982).
[11] Press W. H., Flannery, B. P., Teukolsky, S. A. and Vetterling, W. T., *Numerical Recipes*, Cambridge University Press, Cambridge, 1988.
[12] Choudhary, S. and Felsen, L. B., *IEEE Trans. Antennas Propag.* **AP-21**, 827 (1973).
[13] Mazzucato, E., *Phys. Fluids B* **1**, 1855 (1989).
[14] Nowak, S. and Orefice, A., *Phys. Plasmas* **1**, 1242 (1994).
[15] Farina, D., *Fusion Sci. Technol.* **52**, 154 (2007).
[16] Siegman, A. E., *Lasers*, University Science Books, Mill Valley, CA, 1986.

CHAPTER 4

REFRACTIVE INDEX
MEASUREMENTS

As we have seen in Chapter 2, the propagation of electromagnetic waves in plasmas has a strong dependence on both density and magnetic field. This is why measurements of the plasma refractive index have found extensive use in magnetic fusion research and are expected to play a role in future burning plasma experiments. In this chapter, we shall discuss some of these applications.

4.1 Interferometry

Interferometry is a technique where information on the refractive index of a medium is obtained from the interference of probing waves. The basic scheme of a plasma interferometer [1, 2] is illustrated in Fig. 4.1, where an electromagnetic wave from a coherent source of radiation is split into two parts by a beam-splitter, one used for probing the plasma and the other serving as a reference wave. The two waves are then added together and the power is measured for example by a square law detector. Denoting with $E_r \cos(\omega t)$ the reference wave and with $E_s \cos(\omega t + \Delta\varphi)$ the probing wave (with $\Delta\varphi$ the phase shift induced by the plasma), the output signal (S) is

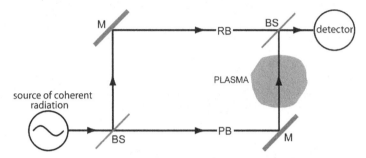

Figure 4.1. Basic scheme of a plasma interferometer using homodyne detection (M: mirror, BS: beam-splitter, PB: probing beam, RB: reference beam).

proportional to

$$S \propto [E_s \cos(\omega t + \Delta\varphi) + E_r \cos(\omega t)]^2$$
$$= \frac{E_s^2}{2}[1 + \cos(2\omega t + 2\Delta\varphi)] + \frac{E_r^2}{2}[1 + \cos(2\omega t)]$$
$$+ E_s E_r [\cos(2\omega t + \Delta\varphi) + \cos\Delta\varphi],$$

which when averaged over a time $\delta t \gg 2\pi/\omega$ gives

$$\bar{S} \propto \frac{E_s^2 + E_r^2}{2} + E_s E_r \cos\Delta\varphi. \tag{4.1}$$

In principle, one could obtain the phase shift induced by the plasma by comparing the measured signals with and without the plasma, which however could be done only when $\Delta\varphi < \pi$. In the following we will discuss how to modify this scheme, but first let us consider what kind of waves should be employed in plasma interferometry.

Obviously, the plasma must be transparent to the probing wave, which implies $\omega > \omega_O$ for the ordinary mode and $\omega > \omega_R$ for the extraordinary mode. However, this is not sufficient for two reasons. The first is that the probing frequency must be large enough for refractive effects (Fig. 4.2) to be negligible. This is a must for burning plasma experiments where plasma accessibility will be very limited. The second reason is that electron cyclotron absorption of the probing wave must be negligible as well. In hot thermonuclear plasmas, this imposes to the frequency of the probing wave to be several times larger than the electron cyclotron frequency [3, 4]. The conclusion is

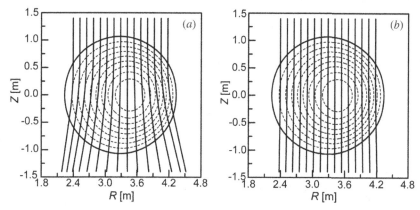

Figure 4.2. Refractive effects in a tokamak plasma with a central electron density of 2.0×10^{20} m^{-3} and a toroidal field of 5 T; (a) wave rays (launched from the top) with frequency of 300 GHz and the O-mode (dash lines are magnetic surfaces); (b) same as (a) with a probing frequency of 1000 GHz.

that the probing wave must be in the submillimeter range of frequencies ($\lambda < 500\,\mu$m), where fortunately several types of far-infrared lasers operate [2, 5, 6]. In this range of frequencies, the wave refractive index in magnetically confined plasmas can be approximated by that of the ordinary mode, i.e., $N = (1 - \omega_p^2/\omega^2)^{1/2} \approx 1 - \omega_p^2/2\omega^2$, so that $\Delta\varphi$ is given by (in Gaussian units)

$$\Delta\varphi = \frac{\omega}{c} \int (N - 1)dl = -\frac{\omega}{2c} \int \frac{\omega_p^2}{\omega^2}dl = -r_e\lambda \int n_e dl, \qquad (4.2)$$

where $r_e = 2.82 \times 10^{-13}$ cm is the classical electron radius and the integral is over the path of the probing beam in the plasma.

We have already mentioned the difficulty of measuring large phase shifts with the interferometer scheme of Fig. 4.1 using homodyne detection [7]. At first sight one might be tempted to suggest that this problem could be overcome by measuring the phase in small steps, starting from the plasma beginning. However, this would not solve the problem since an intrinsic feature of homodyne detection is its inability to determine the sign of the time variation of $\Delta\varphi$. This occurs at the maxima or minima of $\cos\Delta\varphi$ in (4.1), where the sign of the change of measured signals does not depend on the sign of the change in $\Delta\varphi$. We can solve this problem using heterodyne

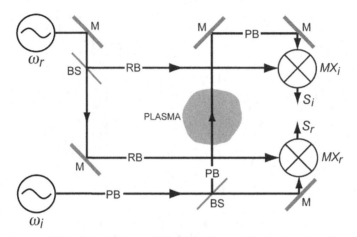

Figure 4.3. Schematic diagram of a plasma interferometer using heterodyne detection (M: mirror; BS: beam-splitter; PB: probing beam; RB: reference beam; MX: mixer).

detection — a well-established technique in the field of radar and radio communications [7].

The schematic diagram of a heterodyne interferometer is illustrated in Fig. 4.3, where the major change from the homodyne scheme is that the reference (E_r) and probing (E_i) waves have different but close frequencies ω_r and ω_i. The two waves are mixed together at two points, one in mixer MX_r before the probing beam has entered the plasma — producing the reference signal S_r — and the other in mixer MX_i after the probing beam has sampled the plasma — producing the interferometer signal S_i. The latter signal is given by

$$S_i \propto [E_i \cos(\omega_i t + \Delta\varphi) + E_r \cos(\omega_r t)]^2$$

$$= \frac{E_i^2 + E_r^2}{2} + \frac{E_i^2}{2}\cos(2\omega_i t + 2\Delta\varphi) + \frac{E_r^2}{2}\cos(2\omega_r t)$$

$$+ E_i E_r [\cos((\omega_i + \omega_r)t + \Delta\varphi)] + E_i E_r [\cos((\omega_i - \omega_r)t + \Delta\varphi)].$$

Then, by filtering out the constant term and those with frequencies $2\omega_i$, $2\omega_r$ and $\omega_i + \omega_r$, we obtain

$$S_i \propto E_i E_r (\cos((\omega_i - \omega_r)t + \Delta\varphi),$$

and similarly for the reference signal S_r

$$S_r \propto E_i E_r (\cos((\omega_i - \omega_r)t)).$$

The advantage of heterodyne detection arises from the fact that, since the time variation of $\Delta\varphi$ is equivalent to a frequency shift $\Delta\omega = d\Delta\varphi(t)/dt$, the sign of the latter can be determined because the intermediate frequency $\omega_i - \omega_r$ is not zero as in homodyne detection.

The two signals are then sent to a phase comparator where the phase $\Delta\varphi$ is obtained using either a digital or an electronic procedure. A common method of the latter type is the quadrature detection system (Fig. 4.4) where the reference signal is split into two parts, with one phase shifted by $90°$. After their mixing with the plasma signal and filtering out the intermediate frequency $(\omega_i - \omega_r)$, one obtains the in-phase signal I and the quadrature signal Q, from which the *phasor* P as defined by

$$P = I + iQ \tag{4.3}$$

is obtained. The phase ϕ of the complex number P satisfies the equation

$$\frac{d\phi}{dt} = \frac{d(\Delta\varphi)}{dt}, \tag{4.4}$$

which yields $\Delta\varphi$ from the measurement of ϕ.

In the noisy environment of fusion experiments, vibrations of optical components may cause spurious variations of the path length, causing phase changes that are inversely proportional to the wavelength of the probing wave. Since the phase shift by the plasma is instead proportional to λ, vibrations can cause serious problems to interferometry measurements using submillimeter or far-infrared waves. Thus, additional information is required for taking this effect into account, which can be obtained using a *two-color interferometry scheme* where two interferometers with different probing wavelengths share the same optical path [8]. The wavelengths (λ_1 and λ_2) of the two probing beams should be chosen as different as possible, with one more sensitive to the plasma and the other to the spurious changes in optical path. If we indicate with $\bar{n}_e$ the average electron density,

L_p the path length in the plasma and ΔL the spurious change in optical path, from (4.2) we get the system of equations in the two unknown $\bar{n}_e$ and ΔL

$$\Delta\varphi_1 = -r_e\lambda_1\bar{n}_eL_p + \frac{2\pi}{\lambda_1}\Delta L,$$

$$\Delta\varphi_2 = -r_e\lambda_2\bar{n}_eL_p + \frac{2\pi}{\lambda_2}\Delta L,$$

(4.5)

yielding

$$\bar{n}_eL_p = \frac{1}{r_e}\frac{\lambda_1\Delta\varphi_1 - \lambda_2\Delta\varphi_2}{\lambda_2^2 - \lambda_1^2}.$$

(4.6)

From this, if $\delta\varphi$ is the diagnostic phase resolution, we get the density resolution

$$\delta\bar{n}_e = \frac{\delta\varphi}{r_eL_p|\lambda_2 - \lambda_1|},$$

(4.7)

which demonstrates the importance of using probing waves with a large wavelength separation.

Even though the line-averaged density is an important parameter, from interferometry measurements one would like to get the local value of the electron density. For cylindrically symmetric plasmas, this can be done quite easily, at least in principle. In this case, if we

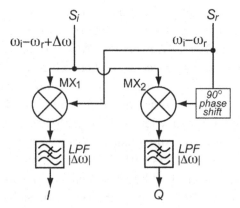

Figure 4.4. Schematic diagram of quadrature detection (S_i: interferometer signal, S_r: reference signal, $\Delta\omega = d\Delta\varphi(t)/dt$, MX: mixer, LPF: low-pass filter).

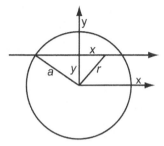

Figure 4.5. Interferometry measurements in a cylindrical plasma.

indicate with f the line integral of n_e, we get (Fig. 4.5)

$$f(y) = \int_{-(a^2-y^2)^{1/2}}^{(a^2-y^2)^{1/2}} n_e dx. \qquad (4.8)$$

With the change of variable $x = (r^2 - y^2)^{1/2}$, this becomes

$$f(y) = 2 \int_y^a n_e(r) \frac{r\, dr}{(r^2 - y^2)^{1/2}}, \qquad (4.9)$$

which can be Abel-inverted to [9]

$$n_e(r) = -\frac{1}{\pi} \int_r^a \frac{df}{dy} \frac{dy}{(y^2 - r^2)^{1/2}} \qquad (4.10)$$

provided $n_e(a) = 0$.

This is a simple expression giving the local plasma density from the line-integrated signals of an interferometer with a sufficient number of channels. Unfortunately, this cannot be applied to the large plasmas of thermonuclear research that are far from being cylindrically symmetric, as Fig. 1.6 clearly demonstrates. In these cases, then, to obtain the local density from interferometer measurements in magnetically confined plasmas one needs to combine the measurements with additional assumptions on the plasma under investigation. One that we can make with great confidence is that the density is constant on magnetic surfaces because of the large electron diffusivity along the magnetic field lines that, as explained in Chapter 1, cover entirely a magnetic surface. This together with the plasma axis-symmetry allows the measurement of the local density on the equatorial plane of a tokamak.

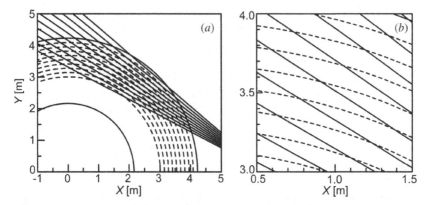

Figure 4.6. Schematic view of a multi-channel interferometer for the measurement of plasma density on the equatorial plane of a tokamak; (a): straight solid
lines are lines of sight of interferometer channels, dash lines are intersections of
magnetic surfaces with the equatorial plane; (b): expanded portion of (a).

To show how, let us consider an interferometer system with n
probing beams having a frequency large enough for their trajectories to be straight lines on the equatorial plane. Let us also choose
n concentric circles (Fig. 4.6), including the outer plasma boundary, with center on the axis of symmetry. Because of the plasma
axis-symmetry, each of these circles coincides with the intersection
of a magnetic surface with the equatorial plane, and is therefore a
line of constant plasma density. Finally, let I and J be two set of
indexes (each running from 1 to n), one (I) labeling the interferometer channels and the other (J) the concentric circles. The trajectory
of channel I is divided into a number of segments of variable length
$\Delta(J_1, J_2)$ by two adjacent crossings of magnetic surfaces J_1 and J_2.
Each segment, then, contributes to the line-integrated density $f(I)$
of channel I by the amount

$$\Delta f(I) = \frac{n_e(J_1) + n_e(J_2)}{2} \Delta(J_1, J_2). \tag{4.11}$$

Adding all terms of this type for a fixed value of I we obtain the
line-integrated density of channel I. By repeating this for all values
of I we obtain a system of n linear equations in the n unknown $n_e(J)$
with constant terms given by the measured line-integrated densities.
The solution of this system of equations yields the local density.

Extending these values of density to the rest of the plasma requires the knowledge of the full 2D-geometry of the magnetic configuration, i.e., we need to know the shape of magnetic surfaces. As a matter of fact, a full knowledge of the plasma equilibrium would allow the use of the above procedure with a set of interferometers having arbitrary line of sights, i.e., using the crossing of the probing beams with an arbitrary set of magnetic surfaces. Indeed, this is the only option for the case of a non-axis-symmetric configuration.

Finally, we must touch the problem of space resolution, a fundamental issue in any diagnostic. Obviously, a good space resolution requires the use of narrow probing beams — with the added fringe benefit of small ports. To see why, let us consider a circular Gaussian beam propagating in vacuum with normalized field amplitude

$$A = \left(\frac{2}{\pi} \right)^{1/2} \frac{1}{w(z)} \exp \left[-\frac{r^2}{w^2(z)} \right], \qquad (4.12)$$

where r is the radial distance from the axis of the beam, z is a coordinate along the direction of propagation and $w(z)$ is the radius where the amplitude drops to $1/e$ of its value on axis. The fractional power transfer of a Gaussian beam through a circular aperture of radius a is given by

$$\frac{2}{\pi w^2} \int_0^a 2\pi r \exp \left[-\frac{2r^2}{w^2} \right] dr = 1 - \exp \left[-\frac{2a^2}{w^2} \right], \qquad (4.13)$$

so that an aperture with $a = w$ transmits $\approx 86\%$ of the total beam power, and $\approx 99\%$ for $a = 2w$.

As described in Chapter 3, the paraxial solution of the Helmholtz equation [10] shows that the radius of a Gaussian beam is given by

$$w(z) = w_0 \left[1 + \left(\frac{z}{z_R} \right)^2 \right]^{1/2}, \qquad (4.14)$$

where $z_R = \pi w_0^2 / \lambda$ is the Rayleigh distance and w_0 is the beam waist — the minimum value of w. From (4.14), we get $w = \sqrt{2} w_0$ at a distance z_R from the waist and that $w(z)$ increases linearly with z for $z \gg z_R$, so that the angular diverge of the beam is $\theta \approx 2\lambda / \pi w_0$. We may define $d = 2z_R$ as the beam depth of focus. As an

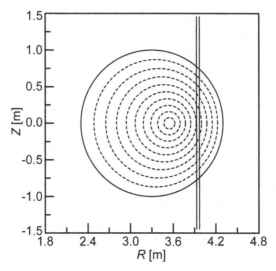

Figure 4.7. Ray tracing of a Gaussian beam with $\lambda = 118.8\,\mu$m and $w_0 = 2\,$cm in the plasma of Fig. 4.2 (using the complex-eikonal method of Chapter 3).

example, for the line of a CH_3OH laser with $\lambda = 118.8\,\mu$m, often used in plasma interferometry [11–13], we get a depth of focus of 21 meters for $w_0 = 2\,$cm. A ray tracing of such a beam in the plasma of Fig. 4.2 is displayed in Fig. 4.7, showing no sign of refractive bending or diffractive divergence. This demonstrates how the use of submillimeter waves can satisfy the demanding needs of both a good spatial resolution and small size ports.

4.2 Polarization of Waves

As we saw in Chapter 2, the polarization of a plasma electromagnetic wave is a strong function of magnetic field. Hence the measurement of changes in wave polarization during plasma propagation can provide important information on the magnetic field. This is very important since there are very few methods for measuring the magnetic field in the interior of hot plasmas.

In Chapter 2, we have also seen that the two characteristic modes of wave propagation in plasmas have elliptical orthogonal polarizations. In a system of Cartesian coordinates (x, y, z) where the wave vector $\mathbf{k}$ is along the z-axis and the magnetic field $\mathbf{B}$ is in the

(x, z)-plane, the polarization $\rho = E_x/E_y$ is given by (2.58) that for convenience we rewrite in the more compact form

$$\rho_\pm = i\frac{1 \pm (1 + F^2)^{1/2}}{F}, \tag{4.15}$$

with

$$F = \frac{2(1 - X)\cos\theta}{Y\sin^2\theta}, \tag{4.16}$$

where θ is the angle between $\mathbf{B}$ and the z-axis and the $\pm$ sign refers to the ordinary $(+)$ and extraordinary $(-)$ modes. Any polarized wave propagating in the z-direction with frequency ω can be written as a unique linear combination of these two orthogonal polarizations. In fact, taking as basis the vectors $(\mathbf{e}_+, \mathbf{e}_-)$ with

$$
\begin{aligned}
e_{+x} &= \rho_+, & e_{+y} &= 1, \\
e_{-x} &= \rho_-, & e_{-y} &= 1,
\end{aligned} \tag{4.17}
$$

a wave propagating in the z-direction with components E_x and E_y can be written as

$$
\begin{aligned}
E_x &= E_+\rho_+ + E_-\rho_-, \\
E_y &= E_+ + E_-,
\end{aligned} \tag{4.18}
$$

with

$$
\begin{aligned}
E_+ &= \frac{E_x - \rho_- E_y}{\rho_+ - \rho_-}, \\
E_- &= \frac{\rho_+ E_y - E_x}{\rho_+ - \rho_-}.
\end{aligned} \tag{4.19}
$$

Let us consider the same wave with Cartesian real components

$$
\begin{aligned}
E_x &= a_x \cos(\tau + \delta_x), \\
E_y &= a_y \cos(\tau + \delta_y),
\end{aligned} \tag{4.20}
$$

where τ denotes the variable part of the phase factor $\mathbf{k}\cdot\mathbf{r} - \omega t$ and δ_x and δ_y are two constant phases. Following [14], these equations are rewritten in the form

$$\frac{E_x}{a_x} = \cos\tau\cos\delta_x - \sin\tau\sin\delta_x,$$

$$\frac{E_y}{a_y} = \cos\tau\cos\delta_y - \sin\tau\sin\delta_y,$$

(4.21)

from which we get

$$\frac{E_x}{a_x}\sin\delta_y - \frac{E_y}{a_y}\sin\delta_x = \cos\tau\sin\delta,$$

$$\frac{E_x}{a_x}\cos\delta_y - \frac{E_y}{a_y}\cos\delta_x = \sin\tau\sin\delta,$$

(4.22)

where $\delta = \delta_y - \delta_x$. Squaring and adding gives

$$\left(\frac{E_x}{a_x}\right)^2 + \left(\frac{E_y}{a_y}\right)^2 - 2\frac{E_x}{a_x}\frac{E_y}{a_y}\cos\delta = \sin^2\delta, \qquad (4.23)$$

which is the equation of an ellipse (Fig. 4.8).

The ellipse is inscribed into a rectangle whose sides are parallel to the coordinates axes and whose lengths are $2a_x$ and $2a_y$. If a and b are the major and minor semi-axes of the ellipse, the state of polarization can be described (Fig. 4.8) by the angle $\psi(0 \leq \psi \leq \pi)$ between the x-axis and the ellipse major semi-axis, and by the angle χ defined by $\tan\chi = \pm b/a \ (-\pi/4 \leq \chi \leq \pi/4)$ with positive or negative sign for

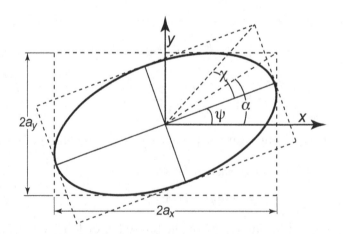

Figure 4.8.   The polarization ellipse.

left-handed or right-handed rotation (here and in the following we will refer to rotations as seen by an observer looking in the direction of wave propagation).

Let (ξ, η) be a new system of coordinates with axes along those of the ellipse. Then the electric field components E_ξ and E_η are

$$E_\xi = E_x \cos\psi + E_y \sin\psi,$$
$$E_\eta = -E_x \sin\psi + E_y \cos\psi. \tag{4.24}$$

If a and b $(a \geq b)$ are the semi-axes of the ellipse, we can also write

$$E_\xi = a\cos(\tau + \delta_0),$$
$$E_\eta = \pm b\sin(\tau + \delta_0), \tag{4.25}$$

where as before the $\pm$ sign indicates the sense in which the electric field end point describes the ellipse. From these equations and (4.21), we get

$$a(\cos\tau\cos\delta_0 - \sin\tau\sin\delta_0) = a_x(\cos\tau\cos\delta_x - \sin\tau\sin\delta_x)\cos\psi$$
$$+ a_y(\cos\tau\cos\delta_y - \sin\tau\sin\delta_y)\sin\psi,$$
$$\pm b(\sin\tau\cos\delta_0 + \cos\tau\sin\delta_0) = -a_x(\cos\tau\cos\delta_x - \sin\tau\sin\delta_x)\sin\psi$$
$$+ a_y(\cos\tau\cos\delta_y - \sin\tau\sin\delta_y)\cos\psi.$$

Equating the coefficient of $\sin\tau$ and $\cos\tau$ yields

$$a\cos\delta_0 = a_x\cos\delta_x\cos\psi + a_y\cos\delta_y\sin\psi, \tag{4.26}$$
$$a\sin\delta_0 = a_x\sin\delta_x\cos\psi + a_y\sin\delta_y\sin\psi, \tag{4.27}$$
$$\pm b\cos\delta_0 = a_x\sin\delta_x\sin\psi - a_y\sin\delta_y\cos\psi, \tag{4.28}$$
$$\pm b\sin\delta_0 = -a_x\cos\delta_x\sin\psi + a_y\cos\delta_y\cos\psi. \tag{4.29}$$

On squaring and adding the first two of these equations we obtain

$$a^2 = a_x^2\cos^2\psi + a_y^2\sin^2\psi + 2a_x a_y \sin\psi\cos\psi\cos\delta,$$

and similarly from the last two equations

$$b^2 = a_x^2\sin^2\psi + a_y^2\cos^2\psi - 2a_x a_y \sin\psi\cos\psi\cos\delta,$$

and therefore

$$a^2 + b^2 = a_x^2 + b_y^2. \tag{4.30}$$

Then multiplying (4.26) by (4.28), (4.27) by (4.29) and adding gives

$$\pm ab = a_x a_y \sin \delta. \tag{4.31}$$

Further, dividing (4.28) by (4.26) and (4.29) by (4.27) gives

$$\pm \frac{b}{a} = \frac{a_x \sin \delta_x \sin \psi - a_y \sin \delta_y \cos \psi}{a_x \cos \delta_x \cos \psi + a_y \cos \delta_y \sin \psi}$$

$$= \frac{-a_x \cos \delta_x \sin \psi + a_y \cos \delta_y \cos \psi}{a_x \sin \delta_x \cos \psi + a_y \sin \delta_y \sin \psi},$$

from which we get

$$(a_x^2 - a_y^2) \sin 2\psi = 2 a_x a_y \cos \delta \cos 2\psi,$$

and therefore

$$\tan 2\psi = \frac{2 a_x a_y}{a_x^2 - a_y^2} \cos \delta. \tag{4.32}$$

Defining α as the angle in the range $0 \leq \alpha \leq \pi/2$ with tangent

$$\tan \alpha = \frac{a_y}{a_x}, \tag{4.33}$$

from the trigonometric identity $\tan 2\alpha = 2 \tan \alpha (1 - \tan^2 \alpha)$ and (4.32), we get

$$\tan 2\psi = \tan 2\alpha \cos \delta. \tag{4.34}$$

Similarly, from (4.30) and (4.31) we have

$$\pm \frac{2ab}{a^2 + b^2} = \frac{2 a_x a_y}{a_x^2 + a_y^2} \sin \delta, \tag{4.34}$$

which using the trigonometry identity $\sin 2\chi = 2 \tan \chi (1 + \tan^2 \chi)$, can be written as

$$\sin 2\chi = \sin 2\alpha \sin \delta. \tag{4.35}$$

In conclusion, if a_x, a_y and the phase δ are given in an arbitrary set of orthogonal coordinates (x, y) and α ($0 \leq \alpha \leq \pi/2$) is given by (4.33), the semi-axes a and b of the polarization ellipse and the angle

ψ $(0 \leq \psi \leq \pi)$ which the major axis makes with the x-axis satisfy the equations

$$a^2 + b^2 = a_x^2 + a_y^2,$$
$$\tan 2\psi = \tan 2\alpha \cos \delta, \qquad (4.36)$$
$$\sin 2\chi = \sin 2\alpha \sin \delta.$$

Since the angles ψ and χ uniquely describe the state of polarization of a completely polarized wave, it is very useful to introduce the Stokes vector [15, 16] **s** with components

$$s_1 = \cos 2\chi \cos 2\psi,$$
$$s_2 = \cos 2\chi \sin 2\psi, \qquad (4.37)$$
$$s_3 = \sin 2\chi,$$

satisfying the condition $s_1^2 + s_2^2 + s_3^2 = 1$.

These results suggest a simple geometrical visualization of the state of polarization by regarding s_1, s_2 and s_3 as the Cartesian coordinates of a point P on a sphere of unit radius — the Poincaré sphere [14] — with longitude 2ψ and latitude 2χ (Fig. 4.9). The points on this sphere represent all possible states of polarization. Since χ is positive or negative according as the polarization is left-handed or right-handed, it follows that left-handed polarizations

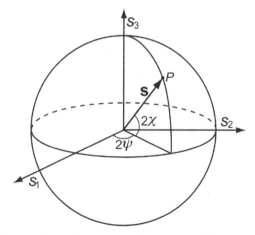

Figure 4.9. Poincaré representation of the state of polarization of a fully polarized wave.

are represented by points lying above the equatorial plane, and right-handed polarizations by those which lie below this plane. Two diametrically opposite points on the sphere represent orthogonal polarizations. Since for linear polarizations the phase difference δ is zero or a multiple of π, the points on the equatorial line represent linear polarizations. Further, the two poles represent the circular polarizations since for these $\chi = \pm\pi/4$.

4.3   Polarization Evolution Equation

In a uniform non-absorbing medium, it can be shown [15–17] that the evolution of the polarization on the Poincaré sphere is represented by a rotation about an axis through the points representing the orthogonal polarizations of the two characteristic modes ($\mathbf{s}_1$ and $\mathbf{s}_2$) by an angle $\Delta\varphi = L(N_1 - N_2)\omega/c$, where L is the path length and N_1 and N_2 are the characteristic refractive indexes (with $N_1 > N_2$). Thus, the evolution of the Stokes vector can be cast in the form

$$\frac{d\mathbf{s}(z)}{dz} = \mathbf{\Omega} \times \mathbf{s}(z), \qquad (4.38)$$

where z is a coordinate along the wave trajectory, and the rotation vector $\mathbf{\Omega}$ has modulus $|\mathbf{\Omega}| = (N_1 - N_2)\omega/c$ and the opposite direction of the fast characteristic mode $\mathbf{s}_2$, so that

$$\mathbf{\Omega} = -\frac{\omega}{c}(N_1 - N_2)\mathbf{s}_2. \qquad (4.39)$$

Equation (4.38) can be applied to a non-uniform medium when gradients are not too large. This is equivalent to assume that each thin slab of thickness dz is approximately uniform.

Following [18–21] and references therein, we discuss the use of (4.38) in the case of wave propagation in magnetized plasmas. We start from (4.15), which gives the polarization of the two characteristic plasma modes — ordinary and extraordinary mode. For the case of interest here, i.e., for a wave frequency much larger than both the plasma frequency and the electron cyclotron frequency, the latter mode is the fast mode. In the plane perpendicular to the direction of wave propagation, from (4.15) we get that the major axis of the polarization ellipse of the X-mode is perpendicular to the component

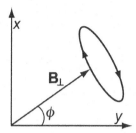

Figure 4.10. Polarization ellipse of the X-mode in the (x,y)-plane of a system of Cartesian coordinates (x, y, z) with the z-axis (pointing into the page) along the direction of wave propagation; $\mathbf{B}_\perp$ is the component of the magnetic field in the (x, y)-plane.

of the magnetic field in the (x, y)-plane (Fig. 4.10). Thus the values of ψ and χ for the fast mode are

$$\psi_2 = \pi - \phi,$$
$$\tan \chi_2 = -[(1 + F^2)^{1/2} - 1]/F, \tag{4.40}$$

where the sign in front of the second expression is negative because $\chi_2 < 0$ for the right-handed polarization of the X-mode. Hence, making use of the trigonometric identity

$$\tan 2\chi_2 = \frac{2 \tan \chi_2}{1 - \tan^2 \chi_2},$$

we get $\tan 2\chi_2 = -F$, giving

$$\sin^2 2\chi_2 = \frac{F^2}{1 + F^2}, \quad \cos^2 2\chi_2 = \frac{1}{1 + F^2},$$

and thus

$$\sin 2\chi_2 = -\frac{F}{(1 + F^2)^{1/2}}, \quad \cos 2\chi_2 = \frac{1}{(1 + F^2)^{1/2}},$$

again with the signs satisfying the constrain $\chi_2 < 0$. From this we obtain

$$\mathbf{s}_2 = \frac{1}{(1 + F^2)^{1/2}} \begin{pmatrix} \cos 2\phi \\ -\sin 2\phi \\ -F \end{pmatrix}. \tag{4.41}$$

To the lowest order in X and Y, the Appleton–Hartree formula (2.32) gives

$$N_1 - N_2 \approx XY \cos\theta, \tag{4.42}$$

so that from (4.39) and (4.41), with $F \approx 2\omega \cos\theta / \omega_c \sin^2\theta$, we obtain

$$\Omega = \frac{\omega_p^2}{2c\omega^3} \begin{pmatrix} -\omega_c^2 \sin^2\theta \cos 2\phi \\ \omega_c^2 \sin^2\theta \sin 2\phi \\ 2\omega\,\omega_c \cos\theta \end{pmatrix}. \tag{4.43}$$

Then, from the two trigonometric identities

$$\sin 2\phi = 2\sin\phi\cos\phi, \quad \cos 2\phi = \cos^2\phi - \sin^2\phi,$$

we get

$$\omega_c^2 \sin^2\theta \cos 2\phi = \left(\frac{e}{mc}\right)^2 B_\perp^2 \cos 2\phi = \left(\frac{e}{mc}\right)^2 (B_y^2 - B_x^2),$$

$$\omega_c^2 \sin^2\theta \sin 2\phi = \left(\frac{e}{mc}\right)^2 B_\perp^2 \sin 2\phi = \left(\frac{e}{mc}\right)^2 2B_x B_y,$$

that together with

$$\omega\,\omega_c \cos\theta = \frac{e}{mc}\omega B_z$$

reduces (4.43) to

$$\Omega = \frac{\omega_p^2}{2c\omega^3} \begin{pmatrix} \left(\frac{e}{mc}\right)^2 (B_x^2 - B_y^2) \\ \left(\frac{e}{mc}\right)^2 2B_x B_y \\ 2\omega\frac{e}{mc}B_z \end{pmatrix}. \tag{4.44}$$

In Gaussian unit, this can be recast in the form

$$\Omega = n_e \begin{pmatrix} C_1(B_x^2 - B_y^2) \\ C_2 B_x B_y \\ C_3 B_z \end{pmatrix}, \tag{4.45}$$

where $C_1 = 2.45 \times 10^{-21}\lambda^3$, $C_2 = 2C_1$, $C_3 = 5.25 \times 10^{-17}\lambda^2$.

The first two components of $\boldsymbol{\Omega}$ define the *Cotton–Mouton* effect, which depends on the component of the magnetic field perpendicular to the direction of wave propagation, while the third is the *Faraday* effect, which instead depends on the parallel component of **B**. An exact expression of $\boldsymbol{\Omega}$ can be found in [21], which does not change the structure of (4.44). On the other hand, the assumption that lead to (4.44), i.e., $X \ll 1$ and $Y \ll 1$, is of vital practical importance for burning plasma experiments, as in the case of interferometry.

4.4 Plasma Polarimetry

Plasma polarimetry is a technique for the measurement of the changes induced by magnetized plasmas on the polarization of a probing wave. As explained in the previous section, the wave polarization can be obtained from (4.36) using the amplitude and phase of two orthogonal components of the wave electric field. In the following, we will first discuss the solution of the polarization evolution equation and then we will briefly describe a polarimeter.

The vector equation (4.38) is the system of three coupled ordinary differential equations

$$\frac{ds_1(z)}{dz} = \Omega_2 s_3 - \Omega_3 s_2,$$

$$\frac{ds_2(z)}{dz} = \Omega_3 s_1 - \Omega_1 s_3, \tag{4.46}$$

$$\frac{ds_3(z)}{dz} = \Omega_1 s_2 - \Omega_2 s_1,$$

which can be written in the compact form

$$\frac{d\mathbf{s}(z)}{dz} = \mathbf{A}(z) \cdot \mathbf{s}(z), \tag{4.47}$$

with **A** the matrix

$$\mathbf{A} = \begin{pmatrix} 0 & -\Omega_3 & \Omega_2 \\ \Omega_3 & 0 & -\Omega_1 \\ -\Omega_2 & \Omega_1 & 0 \end{pmatrix}. \tag{4.48}$$

An approximate solution of (4.47) is given by

$$s_1(z) = s_{01} + s_{03} \int_0^z \Omega_2(z')dz' - s_{02} \int_0^z \Omega_3(z')dz',$$

$$s_2(z) = s_{02} + s_{01} \int_0^z \Omega_3(z')dz' - s_{03} \int_0^z \Omega_1(z')dz', \qquad (4.49)$$

$$s_3(z) = s_{03} + s_{02} \int_0^z \Omega_1(z')dz' - s_{01} \int_0^z \Omega_2(z')dz',$$

where $s_0 = s(0)$, provided

$$\int_0^z |\mathbf{\Omega}|dz' \ll 1. \qquad (4.50)$$

Introducing the functions

$$W_i(z) = \int_0^z \Omega_i dz' \qquad (4.51)$$

and the matrix $\mathbf{M}$

$$\mathbf{M} = \begin{pmatrix} 1 & -W_3 & W_2 \\ W_3 & 1 & -W_1 \\ -W_2 & W_1 & 1 \end{pmatrix} \qquad (4.52)$$

allows to write (4.49) in the compact form

$$s(z) = \mathbf{M}(z) \cdot s_0. \qquad (4.53)$$

From this, we see that the change in polarization depends on both the Cotton–Mouton (W_1 and W_2) and the Faraday (W_3) effects. An interesting case is that with an input linear polarization of $45°$, for which the Stokes vector is $s_0 = (0, 1, 0)$, with the final vector

$$s(z) = (-W_3, 1, W_1) \qquad (4.54)$$

having a polarization rotated by $\Delta\psi = W_3/2$ with respect to the initial $45°$ (Faraday effect), and an ellipticity angle $\chi = W_1/2$ (Cotton–Mutton effect). Interest in this case stems from the non-interference of the two effects, which makes the dependence on density and magnetic field of easier interpretation. For instance for the case of a

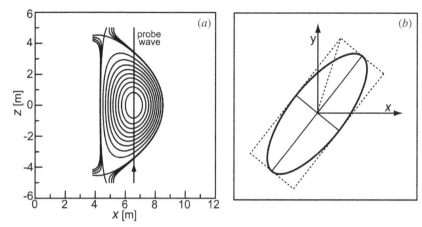

Figure 4.11. Runge–Kutta solution of (4.47) for a probing wave with $\lambda = 500\,\mu m$ and initial linear polarization with $\psi = 45°$; (a): poloidal cross-section of a tokamak plasma (with the y-axis into the page) having a toroidal magnetic field of 6 T and a central density of $2.0 \times 10^{20}\,\text{m}^{-3}$; (b): output polarization with $\psi = 52°$, $\chi = 21°$.

tokamak plasma, if the probing wave is launched perpendicularly to the equatorial plane (as in Fig. 4.11(a)), (4.45) yields

$$W_1 = C_1 \int_0^z n_e(B_x^2 - B_y^2)dz \approx -C_1 B_t^2 \int_0^z n_e dz, \qquad (4.55)$$

where $B_t(= B_y \gg B_x)$ is the toroidal magnetic field, which is almost constant and known along the wave trajectory. Thus, the line average density can be readily obtained from the measurement of χ.

The solution (4.54) of the polarization equation stems from the condition (4.50). When this is not satisfied, one could derive a better approximation by expanding the solution to higher orders [21], or better by solving numerically the polarization equation with the Runge–Kutta method [22]. However, as already mentioned, the plasma accessibility of diagnostics must be kept to a minimum in fusion devices, and thus any technique using electromagnetic waves must minimize their refraction. Moreover, another reason for avoiding refractive effects is that they inevitably complicate the interpretation of measurements. The conclusion then is that the frequency of the probing wave must be much larger than the electron plasma frequency. As shown by (4.44), this implies that the components of the rotation vector are small, and hence the validity of (4.50).

As an example, Fig. 4.11(b) shows the results of the numerical integration of (4.47) for the case of a wave with $\lambda = 500\,\mu$m probing a large size tokamak plasma (Fig. 4.11(a)) with a toroidal magnetic field of 6 T and a large central density of 2×10^{20} m^{-3}. In the system of Cartesian coordinates (x, y, z) where the z-axis is the plasma axis of symmetry, the probing wave propagates with negligible refraction in the vertical direction, starting from an initial linear polarization $(\chi = 0)$ with $\psi = 45°$. The numerical results give $\psi = 52°$ and $\chi = 21°$ for the polarization of the output wave, in excellent agreement with the values of $\psi = 50°$ and $\chi = 21.5°$ that are obtained from (4.54).

Figure 4.12 illustrates the conceptual design of a polarimeter, which is essentially a two-channel heterodyne detection system — very similar to the interferometer scheme of Fig. 4.3. To see how this polarimeter could provide the needed information, we take as an example the case of Fig. 4.11, where as shown in Fig. 4.12 the signal from the plasma is split into two parts and sent to mixers

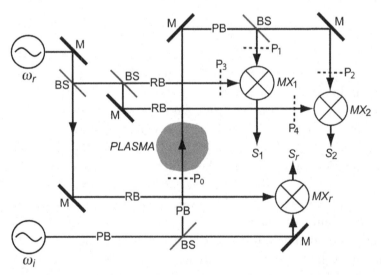

Figure 4.12. Conceptual design of a polarimeter with heterodyne detection (M: mirror; BS: beam-splitter; P: linear polarizer; PB: probing beam; RB: reference beam; MX: mixer).

MX$_1$ and MX$_2$ where they are combined with the reference beam after being filtered by the linear polarizers P_1 and P_2. Each of these polarizers are sitting on a plane perpendicular to the direction of wave propagation, with P_1 being set to maximize in the absence of plasma the signal S_1 from MX$_1$ when the launching polarization is $\chi = 0$ and $\psi = 0$, while P_2 is adjusted to maximizes the signal S_2 from MX$_2$ when the launching polarization is $\chi = 0$ and $\psi = \pi/2$. Under these conditions, the change induced by the plasma on the polarization of a probing wave with initial values $\psi = 45°$ and $\chi = 0$ can be obtained from (4.36) using the amplitudes and phases of measured signals. The latter signals, as in the case of the interferometer scheme of Fig. 4.3, are determined from the comparison of measured signals with the reference signal S_r from mixer MX$_r$.

4.5 Conclusion

Plasma polarimetry is a powerful technique, capable of providing information on the magnetic field of the plasma interior that cannot be obtained from any other plasma diagnostic. From an instrumentation point of view, polarimetry is very similar to interferometry [23–25]. However, while the output of an interferometer can be readily interpreted as the line average plasma density that from a polarimeter is a combination of plasma density and magnetic field. Hence, to get information on one of these parameters, knowledge of the other is needed.

All features of the wave polarization are in the rotation vector $\mathbf{\Omega}$. When the functions W_i of (4.51) are small, the solution of the polarization equation (4.47) can be approximated by (4.53). Inevitably, this is the standard case because of the necessity to use high frequency probing waves for avoiding the deleterious effects of wave refraction. Under these conditions, then, it is possible to obtain the line average density from polarimetry measurements together with other data on the magnetic field that can be used as useful constrains to the reconstruction of magnetohydrodynamic equilibria.

Finally, we must add that in the case of localized magnetic fluctuations, polarimetry measurements of the fluctuating magnetic field are of crucial importance [26].

Bibliography

[1] Heald, M. A. and Wharton, C. B., *Plasma Diagnostic with Microwaves*, Wiley, New York, 1965.

[2] Veron, D., *Infrared and Millimeter Waves*, Vol. 2, Edited by Button, K. J., Academic Press, New York, 1979.

[3] Bornatici, M., Cano, R., De Barbieri, O. and Engelmann, F., *Nucl. Fusion* **23**, 1153 (1983).

[4] Granata, G. and Fidone, I., *J. Plasma Phys.* **45**, 361 (1991).

[5] Mueller, E., Submillimeter Wave Lasers, Wiley Encyclopedia of Electrical and Electronics Engineering, Vol. 20, Edited by Webster, J. G., John Wiley, New York, 1999.

[6] Vasconcellos, E. C. C., Zerbetto, S. C., Zink, L. R., Evenson, K. R. *et al.*, *J. Mol. Spectrosc.* **188**, 102 (1998).

[7] Skolnik, M. I., *Introduction to Radar Systems*, McGraw-Hill, New York, 1962.

[8] Kawano, Y., Nagashima, A., Ishida, S., Fukuda, T. and Matoba, T., *Rev. Sci. Instrum.* **63**, 4971 (1992).

[9] Courant, R. and Hilbert, D., *Methods of Mathematical Physics*, Vol. 1, Interscience Publishers, New York, 1965.

[10] Siegman, A. E., *Lasers*, University Science Books, Mill Valley, CA, 1986.

[11] Wolfe, S. M., Button, K. J., Waldman, J. and Cohn, D. R., *Appl. Opt.* **15**, 2645 (1976).

[12] Mansfield, D. K., Park, H. K., Johnson, L. C. *et al.*, *Appl. Opt.* **26**, 4469 (1987).

[13] Park, H. K., Domier, C. W., Geck, W. R. and Luhmann, N. C., *Rev. Sci. Instrum.* **70**, 710 (1999).

[14] Born, M. and Wolf, E., *Principles of Optics*, Pergamon Press, Oxford, 1965.

[15] Huard, S., *Polarization of Light*, Wiley, New York, 1997.

[16] Goldstein, D. H., *Polarized Light*, CRC Press, New York, 2011.

[17] Ramachandran, G. N. and Ramaseshan S., *Crystal Optics*, Handbuck der Physik, Vol. 35/1, Springer, Berlin, 1961.

[18] DeMarco, F. and Segre, S. E., *Plasma Phys.* **14**, 245 (1972).

[19] Segre, S. E., *Plasma Phys.* **20**, 295 (1978).

[20] Segre, S. E., *Phys. Plasmas*, **2**, 2908 (1995).

[21] Segre, S. E., *Plasma Phys. Control. Fusion* **41**, R57 (1999).

[22] Press W. H., Flannery, B. P., Teukolsky, S. A. and Vetterling, W. T., *Numerical Recipes*, Cambridge University Press, Cambridge, 1988.

[23] Fuchs, C. and Hartfuss, H. J., *Phys. Rev. Lett.* **81**, 1626 (1998).

[24] Brower, D. L., Ding, W. X., Terry, S. D. *et al.*, *Rev. Sci. Instrum.* **74**, (2003).

[25] Petrov, V. G., Petrov, A. A., Malyshev, A. Yu. *et al.*, *Plasma Phys. Rep.* **30**, 111 (2004).

[26] Ding, W. X., Brower, D. L., Terry, S. D. *et al.*, *Phys. Rev. Lett.* **90**, 035002 (2003).

CHAPTER 5

WAVE PROPAGATION IN TURBULENT PLASMAS

As already mentioned in Chapter 1, particle and heat losses in magnetically confined plasmas exceed by several orders of magnitude the predictions of classical collisional diffusion. According to plasma theory [1–3], this is caused by a short-scale turbulence originating from a variety of collective modes which are driven unstable by the inevitable inhomogeneity of confined plasmas — from the ion temperature gradient mode (ITG) and the trapped electron mode (TEM) with the scale of the ion Larmor radius, to the electron temperature gradient mode (ETG) with the scale of the electron Larmor radius. In typical magnetized plasmas of fusion research, we expect the wavelength and frequency of these fluctuations to be in the range of a few centimeters and hundreds of kHz, respectively.

The first evidence of the existence of this type of turbulence in tokamaks was obtained using plasma reflectometry (Chapter 7) [4]. Because of difficulties in getting these results accepted for publication, measurements were quickly repeated using scattering of microwaves [5], which fully confirmed the reflectometry measurements. Since then scattering of electromagnetic waves has played a major role in the study of short-scale turbulence in magnetized plasmas.

In this chapter, we will review the theory of collective scattering of electromagnetic waves by plasma fluctuations and discuss its application to the study of turbulence in fusion plasmas.

5.1 Scattering of Waves by Plasma Fluctuations

Let us consider the case where an electromagnetic wave with electric field

$$\mathbf{E}_i = \mathbf{A}\exp[i(\mathbf{k}_i \cdot \mathbf{r} - \omega_i t)] \tag{5.1}$$

is launched into a plasma occupying the region V of space. For simplicity, we assume that the frequency ω_i is much larger than both the electron plasma frequency ω_p and the electron cyclotron frequency ω_c, so that the wave dielectric tensor becomes the scalar

$$\varepsilon = 1 - \omega_p^2/\omega_i^2. \tag{5.2}$$

In a constant density plasma, the wave by inducing the electric polarization

$$\mathbf{P} = \frac{\varepsilon - 1}{4\pi}\mathbf{E}_i \tag{5.3}$$

changes its phase velocity to $v = c/\sqrt{\varepsilon}$.

Let us now consider the case where the electron density is of the type

$$n(\mathbf{r}, t) = n_0 + n_1(\mathbf{r}, t), \tag{5.4}$$

where $n_1(\mathbf{r}, t)$ is a random function with characteristic frequencies much smaller than ω_i. Then, assuming that the electric polarization is still given by (5.3), the dielectric constant is

$$\varepsilon(\mathbf{r}, t) = 1 - \frac{4\pi n_0 e^2}{m_e \omega_i^2} - \frac{4\pi n_1 e^2}{m_e \omega_i^2} = \varepsilon_0 + \varepsilon_1(\mathbf{r}, t). \tag{5.5}$$

Following [6, 7], we introduce the Hertz vector $\mathbf{\Pi}(\mathbf{r}, t)$ satisfying the wave equation

$$\nabla^2 \mathbf{\Pi} - \varepsilon_0/c^2 \ddot{\mathbf{\Pi}} = -\varepsilon_1 \mathbf{E}, \tag{5.6}$$

where dots indicate the time derivative, with the electric and magnetic fields given by

$$\mathbf{E} = \frac{1}{\varepsilon_0}[\nabla \times \nabla \times \mathbf{\Pi} + \nabla^2 \mathbf{\Pi}] - \frac{1}{c^2}\ddot{\mathbf{\Pi}},$$
$$\mathbf{B} = \tfrac{1}{c}\nabla \times \dot{\mathbf{\Pi}}. \tag{5.7}$$

When $|\varepsilon_1| \ll 1$, (5.6) can be solved with the method of small perturbations where the solution is sought in the form of the series

$$\boldsymbol{\Pi} = \boldsymbol{\Pi}_0 + \boldsymbol{\Pi}_1 + \boldsymbol{\Pi}_2 + \cdots \tag{5.8}$$

with the nth term of order ε_1^n. By substituting (5.8) in (5.6) and equating terms of the same order, we obtain

$$\nabla^2 \boldsymbol{\Pi}_0 - \varepsilon_0/c^2 \ddot{\boldsymbol{\Pi}}_0 = 0,$$
$$\nabla^2 \boldsymbol{\Pi}_1 - \varepsilon_0/c^2 \ddot{\boldsymbol{\Pi}}_1 = -4\pi \mathbf{P}_1, \tag{5.9}$$

where $\mathbf{P}_1 = \varepsilon_1 \mathbf{E}_i/4\pi$.

The first of (5.9) is the wave equation in a homogenous and isotropic medium with dielectric constant ε_0, which for simplicity we will assume equal to 1. The solution of the second is [8]

$$\boldsymbol{\Pi}_1 = \int_V \frac{\mathbf{P}_1(\mathbf{r}', t - R/c)}{R} d\mathbf{r}', \tag{5.10}$$

where $R = |\mathbf{r}' - \mathbf{r}|$, showing that the first approximation of the scattered wave is the radiation from the dipoles induced by the electric field of the incident wave.

Outside of the region V, the electric and magnetic fields are

$$\mathbf{E}_1(\mathbf{r}, t) = \nabla \times \nabla \times \boldsymbol{\Pi}_1,$$
$$\mathbf{B}_1(\mathbf{r}, t) = \frac{1}{c} \nabla \times \dot{\boldsymbol{\Pi}}_1. \tag{5.11}$$

From (5.10), we get

$$\nabla \times \boldsymbol{\Pi}_1 = \int_V d\mathbf{r}' \left\{ \nabla \frac{1}{R} \times [\mathbf{P}_1] + \frac{1}{R} \nabla \times [\mathbf{P}_1] \right\}$$
$$= \int_V d\mathbf{r}' \left\{ -\frac{1}{R^3} \mathbf{R} \times [\mathbf{P}_1] - \frac{1}{R^2 c} \mathbf{R} \times [\dot{\mathbf{P}}_1] \right\}, \tag{5.12}$$

where the square brackets indicate that t must be replaced by the retarded time $t - R/c$. From this, when $R \gg c|\mathbf{P}_1|/|\dot{\mathbf{P}}_1| \approx c/\omega_i$, we get

$$\nabla \times \boldsymbol{\Pi}_1 \approx \int_V d\mathbf{r}' \frac{[\dot{\mathbf{P}}_1] \times \mathbf{R}}{R^2 c}. \tag{5.13}$$

Similarly, for $R \gg c|\dot{\mathbf{P}}_1|/|\ddot{\mathbf{P}}_1| \approx c/\omega_i$, we also get

$$\nabla \times \nabla \times \mathbf{\Pi}_1 = \int_V d\mathbf{r}' \frac{([\ddot{\mathbf{P}}] \times \mathbf{R}) \times \mathbf{R}}{R^3 c^2} \qquad (5.14)$$

and

$$\nabla \times \dot{\mathbf{\Pi}}_1 \approx \int_V d\mathbf{r}' \frac{[\ddot{\mathbf{P}}_1] \times \mathbf{R}}{R^2 c} \qquad (5.15)$$

yielding $\mathbf{E}_1$ and $\mathbf{B}_1$ from (5.11). For $\mathbf{B}_1$ we get

$$\mathbf{B}_1(\mathbf{r}, t) = \frac{e^2}{mc^2} \mathbf{A}$$
$$\times \int_V d\mathbf{r}' \frac{\mathbf{R}}{R^2} n_1 \left(\mathbf{r}', t - \frac{R}{c}\right) \exp\left[i\left(\mathbf{k}_i \cdot \mathbf{r}' - \omega_i \left(t - \frac{R}{c}\right)\right)\right]. \qquad (5.16)$$

By expanding $R = |\mathbf{r}' - \mathbf{r}|$ in series of powers of r'/r, to first order we obtain

$$R \approx r - \mathbf{s} \cdot \mathbf{r}' + \frac{1}{2r} \left[r'^2 - (\mathbf{s} \cdot \mathbf{r}')^2\right], \qquad (5.17)$$

where $\mathbf{s} = \mathbf{r}/r$. If $2r/k_i \gg L^2$, with L the linear size of the scattering region, we can approximate the exponential in (5.16) with

$$\exp[i(k_i r + \mathbf{k}_i \cdot \mathbf{r}' - k_i \mathbf{s} \cdot \mathbf{r}' - \omega_i t)]$$

and replace R with r in the other members of (5.16). Finally, by defining the vector $\mathbf{k}$

$$\mathbf{k} = \mathbf{k}_s - \mathbf{k}_i, \qquad (5.18)$$

where $\mathbf{k}_s = k_i \mathbf{s}$, (5.16) can be cast in the form

$$\mathbf{B}_1(\mathbf{r}, t) = \frac{e^2}{m_e c^2} \frac{\exp[ik_i r]}{r} \mathbf{A} \times \mathbf{s}$$
$$\times \int_V d\mathbf{r}' n_1 \left(\mathbf{r}', t - \frac{r}{c}\right) \exp[-i(\mathbf{k} \cdot \mathbf{r}' + \omega_i t)]. \qquad (5.19)$$

5.2 Intensity of Scattered Waves

The energy flux of scattered waves is

$$\mathbf{F}(\mathbf{r}, t) = \frac{c}{4\pi} \operatorname{Re} \mathbf{E}_1 \times \operatorname{Re} \mathbf{B}_1 = \frac{c}{4\pi} (\operatorname{Re} B_1)^2 \mathbf{s}, \qquad (5.20)$$

where Re indicates the real component. However, what has physical meaning is only the average of $\mathbf{F}$ over a time interval that is longer than both the period of the incident wave and the correlation time of plasma fluctuations. The spectral density of this quantity can be obtained from the Wiener–Khinchin theorem [6], stating that the spectral power density of a random function is the Fourier transform of its auto-correlation function. Thus, calling $\mathbf{F}_\omega$ the spectral density of $\mathbf{F}$, we have

$$\mathbf{F}_\omega(\mathbf{r}) = \mathbf{s} \frac{c}{2\pi} \int_{-\infty}^{+\infty} d\tau \cos(\omega\tau) \langle \operatorname{Re} B_1(\mathbf{r}, t) \operatorname{Re} B_1(\mathbf{r}, t + \tau) \rangle, \quad (5.21)$$

which includes the contribution of both positive and negative frequencies. The function

$$\langle \operatorname{Re} B_1(\mathbf{r}, t) \operatorname{Re} B_1(\mathbf{r}, t + \tau) \rangle, \qquad (5.22)$$

where the brackets indicating the time average, is the time auto-correlation function of $\operatorname{Re} B_1(\mathbf{r}, t)$. By substituting (5.19) into (5.21) we have

$$\mathbf{F}_\omega(\mathbf{r}) = \mathbf{s} \frac{c}{2\pi} \left(\frac{e^2}{m_e c^2} \right)^2 \frac{|\mathbf{A} \times \mathbf{s}|^2}{r^2} \int_{-\infty}^{\infty} d\tau \cos(\omega\tau)$$

$$\times \int_V \int_V d\mathbf{r}_1' d\mathbf{r}_2' \langle n_1(\mathbf{r}_1', t) n_1(\mathbf{r}_2', t + \tau) \cos(k_i r - \mathbf{k} \cdot \mathbf{r}_1' - \omega_i t)$$

$$\times \cos(k_i r - \mathbf{k} \cdot \mathbf{r}_2' - \omega_i(t + \tau)) \rangle. \qquad (5.23)$$

where $|\mathbf{A} \times \mathbf{s}|^2 = A^2$. From this we get the scattering cross-section per unit frequency and solid angle

$$\sigma = \frac{r^2 F_\omega}{(c/8\pi) A^2}$$

$$= 4\sigma_0 \int_{-\infty}^{\infty} d\tau \cos(\omega\tau) \frac{1}{2} \int_V \int_V d\mathbf{r}_1' d\mathbf{r}_2' \langle n_1(\mathbf{r}_1', t) n_1(\mathbf{r}_2', t + \tau)$$

$$\times [\cos[2k_i r - \mathbf{k} \cdot (\mathbf{r}_1' + \mathbf{r}_2') - \omega_i(2t + \tau)]$$
$$+ \cos[-\mathbf{k} \cdot (\mathbf{r}_2' - \mathbf{r}_1') - \omega_i \tau]\}, \tag{5.24}$$

where $\sigma_0 = (e^2/m_e c^2)^2$. The first term under the averaging brackets vanishes and (5.24) becomes

$$\sigma = 2\sigma_0 \int_{-\infty}^{+\infty} d\tau \cos(\omega\tau)$$

$$\times \int_V \int_V d\mathbf{r}_1' d\mathbf{r}_2' \langle n_1(\mathbf{r}_1', t) n_1(\mathbf{r}_2', t + \tau) \rangle \cos[-\mathbf{k} \cdot (\mathbf{r}_2' - \mathbf{r}_1') - \omega_i \tau], \tag{5.25}$$

where

$$\langle n_1(\mathbf{r}_1', t) n_1(\mathbf{r}_2', t + \tau) \rangle \tag{5.26}$$

is the density autocorrelation function, which when the system is homogeneous is an even function of both $\mathbf{r}_2' - \mathbf{r}_1'$ and τ. For such a case, (5.25) becomes

$$\sigma = \sigma_0 \int_{-\infty}^{+\infty} d\tau \{\exp[i(\omega - \omega_i)\tau] + \exp[-i(\omega + \omega_i)\tau]\}$$

$$\times \int_V \int_V d\mathbf{r}_1' d\mathbf{r}_2' \langle n_1(\mathbf{r}_1', t) n_1(\mathbf{r}_2', t + \tau) \rangle \exp[-i\mathbf{k} \cdot (\mathbf{r}_2' - \mathbf{r}_1')]. \tag{5.27}$$

Let us now consider the function $f(\tau)$ as defined by

$$f(\tau) = \int_V \int_V d\mathbf{r}_1' d\mathbf{r}_2' \langle n_1(\mathbf{r}_1', t) n_1(\mathbf{r}_2', t + \tau) \rangle \exp[-i\mathbf{k} \cdot (\mathbf{r}_2' - \mathbf{r}_1')] \tag{5.28}$$

and introduce the new variables

$$\boldsymbol{\rho} = \mathbf{r}_2' - \mathbf{r}_1', \quad \boldsymbol{\eta} = \frac{\mathbf{r}_2' + \mathbf{r}_1'}{2}.$$

By substituting in (5.28) the density autocorrelation function with its Fourier expansion

$$\langle n_1(\mathbf{r}_1', t) n_1(\mathbf{r}_2', t + \tau) \rangle = \frac{1}{8\pi^3} \int d\boldsymbol{\kappa} \, S(\boldsymbol{\kappa}, \tau) \exp(i\boldsymbol{\kappa} \cdot \boldsymbol{\rho}), \tag{5.29}$$

where the integral is over the full domain of κ, and by integrating with respect to η (which gives the volume V) we obtain

$$f(\tau) = \frac{V}{8\pi^3} \int d\kappa\, S(\kappa, \tau) \int_V d\boldsymbol{\rho} \exp[i(\kappa - \mathbf{k}) \cdot \boldsymbol{\rho}]. \qquad (5.30)$$

In this equation, the function,

$$F(\kappa - \mathbf{k}) = \frac{1}{8\pi^3} \int_V d\boldsymbol{\rho} \exp[i(\kappa - \mathbf{k}) \cdot \boldsymbol{\rho}] \qquad (5.31)$$

becomes the Dirac delta function $\delta(\kappa - \mathbf{k})$ when the domain of integration is infinite. For a finite domain, $F(\kappa - \mathbf{k})$ has a maximum at $\kappa = \mathbf{k}$ and falls rapidly to a small value when $|\kappa - \mathbf{k}| > 2\pi/L$ (where L is the minimum linear size of the domain of integration). In this case then we have

$$\int d\kappa\, F(\kappa - \mathbf{k}) \approx \int_V d\boldsymbol{\rho}\, \delta(\boldsymbol{\rho}) = 1. \qquad (5.32)$$

Thus, if $S(\kappa, \tau)$ does not vary appreciably within a region of the wave vector space of volume $8\pi/L^3$, i.e., the correlation distance of plasma fluctuations is much smaller than the linear dimension of the scattering region, a good approximation of (5.27) is

$$\sigma = L^3 \sigma_0 \int_{-\infty}^{+\infty} d\tau \{\exp[i(\omega - \omega_i)\tau] + \exp[-i(\omega + \omega_i)\tau]\} S(\mathbf{k}, \tau).$$
$$(5.33)$$

By indicating with $S(\mathbf{k}, \omega)$ the Fourier transform of $S(\mathbf{k}, \tau)$ with respect to τ, (5.33) can be written as

$$\sigma = L^3 \sigma_0 [S(\mathbf{k}, \omega - \omega_i) + S(\mathbf{k}, -\omega - \omega_i)]. \qquad (5.34)$$

Since the period of the probing wave must be much smaller than the correlation time of fluctuations, the second term in (5.34) is zero, and therefore the final expression for the effective cross-section of the scattering process per unit volume, frequency and solid angle is

$$\sigma = \sigma_0 S(\mathbf{k}, \omega_s - \omega_i) = \sigma_0 S(\mathbf{k}, \omega), \qquad (5.35)$$

where ω in (5.34) has been renamed ω_s to indicate that it is the frequency of scattered waves, while $\omega = \omega_s - \omega_i$ is reserved for the

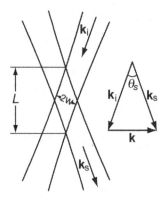

Figure 5.1.   Scattering geometry (left) and Bragg condition (right).

frequency of turbulence. The function $S(\mathbf{k}, \omega)$ is usually referred to as the power spectrum of density fluctuations.

5.3   Turbulence Measurements

Equation (5.35) is the basis for the investigation of plasma turbulence with scattering of electromagnetic waves, where the scattered power at different angles with respect to the probing wave yields the power spectrum $S(\mathbf{k}, \omega)$ of turbulent fluctuations at $\omega = \omega_s - \omega_i$ and $\mathbf{k} = \mathbf{k}_s - \mathbf{k}_i$. These are the Bragg conditions, imposing energy and momentum conservation to the scattering process. Since for the topic of this chapter $\omega_s \approx \omega_i$ and $k_s \approx k_i$, the second of the Bragg conditions gives

$$k = 2k_i \sin\theta_s, \tag{5.36}$$

where θ_s is the scattering angle (Fig. 5.1).

The instrumental resolution of scattering measurements is limited by the size of the probing beam and the radiation pattern of the receiving antenna. If for both we assume the Gaussian amplitude profile $A(r_\perp) = \exp(-r_\perp^2/w^2)$, where $r_\perp$ is the radial coordinate perpendicular to the direction of propagation and w is the beam waist (Fig. 5.1), the wave number resolution of measured fluctuations, which is obtained from the Fourier expansion of $A(r_\perp)$, is given by $G(\kappa_\perp) = \exp(-\kappa_\perp^2/\Delta^2)$, where $\Delta = 2/w$ and $\kappa_\perp$ is the beam wave number perpendicularly to the direction of propagation.

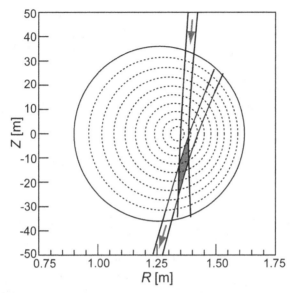

Figure 5.2. Scattering geometry for detection of turbulent fluctuations with $k = 7\,\mathrm{cm}^{-1}$ in the Princeton Large Torus (PLT) tokamak [10] with $\omega_i/2\pi = 140\,\mathrm{GHz}$.

If we take the average linear dimension of the common region between probing and scattered beams as a measure of spatial resolution, we get $L/2 = w/\sin(\theta_s/2) = 2wk_i/k$, which for the case of Fig. 5.2 (with $w = 2\,\mathrm{cm}$, $k_i = 29\,\mathrm{cm}^{-1}$ and $k = 7\,\mathrm{cm}^{-1}$) gives 16 cm.

From (5.35), the ratio of scattered (P_s) to incident (P_i) power is

$$\frac{P_s}{P_i} = \sigma_0 S(\mathbf{k}, \omega) L \Omega, \tag{5.37}$$

where $\Omega = \pi(2/k_i w)^2$ is the solid angle of the receiving antenna. For an estimate of P_s, we replace the power spectrum S in (5.37) with its average value $\langle S \rangle$ that can be obtained from

$$\langle \delta n^2 \rangle = \frac{\langle S \rangle}{(2\pi)^3}(\pi k_\perp^2)\frac{2\pi}{L}, \tag{5.38}$$

where $\langle \delta n^2 \rangle$ is the mean square density fluctuation and $k_\perp$ is the average wave number of fluctuations perpendicularly to the magnetic field. From this we get

$$\frac{P_s}{P_i} = \left(\frac{\omega_p}{\omega_i}\right)^4 \left(\frac{k_i L}{k_\perp w}\right)^2 \frac{\langle \delta n^2 \rangle}{n^2}, \tag{5.39}$$

showing the dependence of the scattered power on plasma and scattering parameters. For the case of Fig. 5.2 with $\omega_p/\omega_i = 10^{-1}$, $k_\perp = 7\,\mathrm{cm}^{-1}$ and $\langle \delta n^2 \rangle / n^2 = 10^{-4}$, we get $P_s/P_i = 1 \times 10^{-5}$.

Since for the correct interpretation of measurements we need the sign of $\omega = \omega_s - \omega_i$, heterodyne detection must be employed in scattering experiments. For this, we can use the scheme of Fig. 4.3 with the obvious change that now it is the scattered signal — not the probing wave — that must be sent to the mixer MX_1. Quadrature detection must be used as well, with the final output consisting in the fast fourier transform of the phasor P of (4.3).

An example of the measured spectrum of fluctuations using the scattering geometry of Fig. 5.2 together with heterodyne detection is displayed in Fig. 5.3, showing a spectrum that is clearly shifted towards the side of negative frequencies. From this and the knowledge of $\mathbf{k} = \mathbf{k}_s - \mathbf{k}_i$, we conclude that the measured fluctuations propagate along the electron diamagnetic velocity ($\mathbf{v}_{\mathrm{De}} = \nabla p_e \times \mathbf{B}/en_e B^2$, with ∇p_e the electron pressure gradient). This is an important piece of information since it indicates that these fluctuations are driven by electrons [3].

Another indication that the frequency shift shown in Fig. 5.3 is due to a phase propagation along the electron diamagnetic velocity is in Fig. 5.4, which shows that the sign of the frequency shift changes when the direction of the toroidal magnetic field is reversed, making the phase velocity along v_{De} to correspond to positive frequencies in case (a) and negative frequencies in case (b).

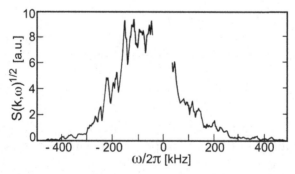

Figure 5.3.   Power spectrum of plasma fluctuations using the scattering geometry of Fig. 5.2. (With permission from [10]).

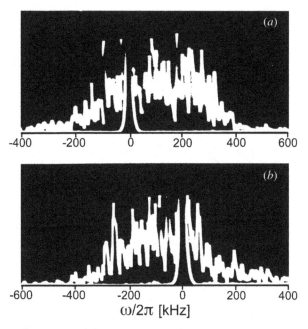

Figure 5.4.   Spectrum of fluctuations similar to those of Fig. 5.3 in plasmas with the toroidal magnetic field in opposite directions.

In tokamaks, it is well know that the plasma torus can rotate in the azimuthal direction because of either an outside injection of momentum or the onset of a spontaneous rotation, the latter being a phenomenon still not well understood. To explain how this could modify the spectrum of small-scale fluctuations when observed in the laboratory frame, let us consider the orthogonal system of coordinates (θ, ϕ, ψ) of Fig. 5.5, where the unit vector $\mathbf{e}_\psi$ is parallel to the outward normal to the magnetic surface (i.e., $\nabla p \cdot \mathbf{e}_\psi < 0$, where ∇p is the plasma pressure gradient), $\mathbf{e}_\phi$ is parallel to the toroidal plasma current and $\mathbf{e}_\theta = \mathbf{e}_\phi \times \mathbf{e}_\psi$ (i.e., $B_\theta > 0$). Since the amplitude of microinstabilities in tokamak plasmas is almost constant along the magnetic field lines [3], we may assume that the wave vector of measured fluctuations is parallel to the plane $(\mathbf{e}_\theta, \mathbf{e}_\psi)$. Let us then refer to fluctuations that in the plasma frame propagate along the electron diamagnetic velocity $\mathbf{v}_{De}$ as electron waves, and those propagating along the ion diamagnetic velocity $\mathbf{v}_{Di} = -\nabla p_i \times \mathbf{B}/en_i B^2$ as ion

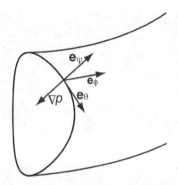

Figure 5.5.   Orthogonal system of coordinate (θ, φ, ψ).

waves. Then, from the equations

$$\mathbf{v}_{\text{De}} \cdot \mathbf{e}_\phi = -\frac{|\nabla p_e| B_\theta}{e n_e B^2} < 0,$$

$$\mathbf{v}_{\text{Di}} \cdot \mathbf{e}_\phi = \frac{|\nabla p_i| B_\theta}{e n_i B^2} > 0,$$

we conclude that the Doppler effect from a plasma co-rotation (i.e., in the plasma current direction) is that of shifting the measured scattering spectrum of electron waves towards the ion side, while a counter-rotation would obviously do the opposite. For ion waves, co- and counter-rotations reverse their role, i.e., a counter-rotation shifts the measured scattering spectrum of ion waves towards the electron side, while a co-rotation does the opposite.

In conclusion, if the nature of observed fluctuations is known, the measured Doppler shift provides information on plasma rotation. On the other hand, if the latter is know from other measurements, the observed dependence of the measured Doppler shift on plasma rotation provides information on the propagation of fluctuations in the plasma frame.

Measurements of turbulent fluctuations with scattering of electromagnetic waves in tokamaks can be divided into two groups, the first employing millimeter waves (75–300 GHz) [5, 10–20], and the second using CO_2 lasers (3×10^4 GHz) [21–27]. There are advantages and disadvantages in both of these approaches. The power of available CW CO_2 lasers is in the range of 25–50 W, which is more than three orders of magnitude larger than what is available for millimeter waves. On the other hand, the scattering angles for CO_2 lasers are much smaller than for millimeter waves leading to a poor spatial resolution and to low sensitivity to density fluctuations. This is because, even though the scattering cross-section (5.35) does not depend on the wavelength (λ) of the probing wave, the small scattering angle ($\theta_s \propto \lambda$) makes the collection solid angle small as well ($\Omega \propto \lambda^2$) and thus lowers the power of scattered waves. This is why experiments with millimeter waves are more numerous than those using CO_2 lasers. However, the larger scattering angles of millimeter waves require a larger plasma accessibility as well, which will be very difficult — if not impossible — to be available in the next generation of burning plasma experiments, where plasma accessibility will require penetration of not only toroidal magnets and vacuum vessels, as in present experiments, but also of thick insulating layers and radiation shields. This is why only the use of infrared sources will be feasible for scattering measurements. In the next section, we will see how the anisotropy of tokamak turbulence can help in reducing the size of the scattering volume.

5.4 Short-Scale Anisotropic Turbulence

The estimate of the scattering volume in the previous section is valid only for the case of isotropic turbulence. However, as already mentioned, the short-scale turbulence of tokamak plasmas is anisotropic since, while the scale of turbulent fluctuations perpendicularly to the magnetic field is of the order of ion or electron Larmor radii, along the field lines it is $\sim qR$ (with q the magnetic safety factor and R the plasma major radius). Therefore, for all practical purposes we can assume $\mathbf{k} \cdot \mathbf{B} = 0$. In the following, we will impose this constrain on the range of possible fluctuations, i.e., we will assume the wave vector of fluctuations to be perpendicular to the magnetic field. In this case, the spatial variation of the direction of the magnetic field can modify the instrumental response by detuning the scattering receiver. This can be easily understood when the probing wave propagates perpendicularly to the magnetic surfaces and scattering angles are small (Fig. 5.6). From the beam spectrum $G(\kappa_\perp)$ of the previous section,

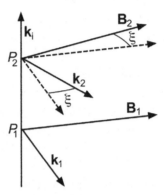

Figure 5.6. Magnetic field $(\mathbf{B}_1, \mathbf{B}_2)$ and fluctuation wave vectors $(\mathbf{k}_1, \mathbf{k}_2)$ at two points (P_1, P_2) of a probing beam propagating perpendicularly to magnetic surfaces. Scattering angles are equal $(k_1 = k_2)$ and small $(\mathbf{k}_1, \mathbf{k}_2$ are nearly perpendicular to $\mathbf{k}_i$). (With permission from [28]).

one can readily obtain the instrumental selectivity function [24], as defined by the collection efficiency of the receiver

$$F(\mathbf{r}) = \exp[-(2k\sin(\xi(r)/2)/\Delta)^2], \tag{5.40}$$

where k is the wave number of fluctuations and $\xi(r)$ is the change in pitch angles of magnetic field lines starting from the point (P_1) where the scattered waves are detected with maximum efficiency, i.e., from the aiming point of the receiving antenna. For $\xi(r) \ll 1$, this yields the spatial resolution $\delta l \approx 2\Delta/k\langle d\xi/dr\rangle$, where $\langle d\xi/dr\rangle$ is the average derivative of the magnetic pitch angle inside the scattering region. Compared with the estimate in the previous section, δl does not depend on the wave number of the probing wave, which is very advantageous for scattering of infrared waves. However, to take full advantage of the magnetic field spatial variation one needs to consider a propagation of the probing beam at an arbitrary angle with the magnetic field [28]. This is because the Bragg condition becomes strongly dependent on the toroidal curvature of magnetic field lines when the probing beam forms a small angle with the magnetic field. This is schematically illustrated in Fig. 5.7 showing a case where both the probing beam and the scattered wave propagate on the torus mid-plane. For a given fluctuation wave number, the wave vector $\mathbf{k}_s$ is perpendicular to a magnetic surface and parallel to

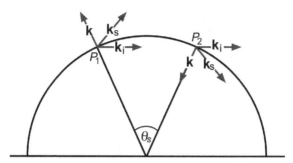

Figure 5.7. Scattering geometry with both probing and scattered waves on the torus mid-plane.

the mid-plane only if the wave is scattered at one of the two points P_1 and P_2 (toroidally separated by an angle equal to the scattering angle), where the fluctuation wave vector $\mathbf{k}$ is in the plasma radial direction. For all of the other probing locations, the Bragg condition imposes to the scattered wave to propagate at an angle with the mid-plane. It is this phenomenon that can be exploited for reducing the size of the scattering region.

To see why, let us consider the system of orthogonal coordinates (u, v, t) with the t-axis parallel to $\mathbf{k}_i$, and define the polar angle φ with

$$k_{su} = k_i \sin\theta_s \cos\varphi, \quad k_{sv} = k_i \sin\theta_s \sin\varphi, \quad k_{st} = k_i \cos\theta_s. \quad (5.41)$$

Let us then consider scattered waves originating from the two points (O_1 and O_2) of the probing beam with identical scattering angles but different wave vectors $\mathbf{k}_s^1$ and $\mathbf{k}_s^2$ (Fig. 5.8). From (5.41) we obtain

$$\frac{\mathbf{k}_s^1 \cdot \mathbf{k}_s^2}{k_i^2} \equiv \cos\alpha = \cos^2\theta_s + \sin^2\theta_s \cos\delta\varphi,$$

where $\delta\varphi = \varphi_2 - \varphi_1$, giving

$$\cos\alpha = 1 - \sin^2\theta_s(1 - \cos\delta\varphi) = 1 - 2\sin^2(\delta\varphi/2)\sin^2\theta_s$$

For $\theta_s^2 \ll 1$ (always satisfied for infrared probing waves), this becomes

$$\alpha^2 \approx 4\theta_s^2 \sin^2(\delta\varphi/2). \quad (5.42)$$

If we assume that the receiving antenna is positioned for collecting with maximum efficiency the scattered waves from O_1, by replacing $\kappa_\perp$ with $k_i\alpha$ in the spectrum $G(\kappa_\perp)$ we obtain the collection efficiency

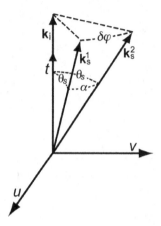

Figure 5.8.   Orthogonal coordinates (u, v, t). (With permission from [29]).

of scattered waves from O_2

$$F = \exp(-\alpha^2/\alpha_0^2)$$

where $\alpha_0 = \Delta/k_i$. This together with the Bragg condition ($k \approx k_i\theta_s$) yields the instrumental selectivity function

$$F = \exp[-(2k\sin(\delta\varphi/2)/\Delta)^2]. \tag{5.43}$$

When the probing beam propagates perpendicularly to the magnetic surfaces, φ becomes the magnetic pitch angle (apart from an additional constant), which makes (5.43) to coincide with (5.40).

The final step is to get the angle φ from the equation

$$(\mathbf{k}_s - \mathbf{k}_i) \cdot \mathbf{B} = \mathbf{k} \cdot \mathbf{B} = 0,$$

which can be written as

$$B_t(\cos\theta_s - 1) + B_u \sin\theta_s \cos\varphi + B_v \sin\theta_s \sin\varphi = 0 \tag{5.44}$$

giving

$$\cos\varphi = \frac{1}{B_\perp^2 \sin\theta_s}\{B_u B_t(1 - \cos\theta_s) \pm [B_u^2 B_t^2(1 - \cos\theta_s)^2$$
$$- B_\perp^2(B_t^2(1 - \cos\theta_s)^2 - B_v^2 \sin^2\theta_s)]^{1/2}\} \tag{5.45}$$

and

$$\sin(\varphi) = \frac{B_t(1 - \cos\theta_s) - B_u \sin\theta_s \cos\varphi}{B_v \sin\theta_s}, \tag{5.46}$$

where $B_\perp^2 = B_u^2 + B_v^2$ and the $\pm$ sign corresponds to the two scattering branches of Fig. 5.7.

5.5 CO_2 Laser Scattering

In this section, the above equations are used for assessing the degree of localization of CO_2 scattering measurements in the next generation of burning plasma experiments [29]. As an example, we use a tokamak plasma with the predicted parameters of ITER [30]: major radius 5.2 m, minor radius 2 m, toroidal magnetic field 5.3 T, elongation 1.8, plasma current 15 MA and central electron density $1.0 \times 10^{20}\,\mathrm{m}^{-3}$.

We begin with the scattering geometry of Fig. 5.9, where a CO_2 probing beam with a waist (w) of 3 cm propagates on the torus equatorial plane along the x-axis, and the scattering receiver is set to measure fluctuations with wave vectors parallel to the equatorial plane. Here and in the following we will use the system of

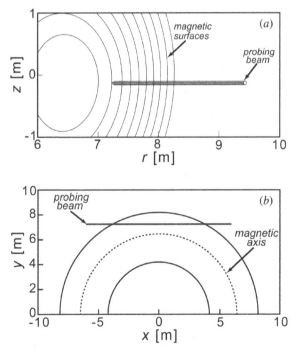

Figure 5.9. Poloidal (a) and toroidal (b) ray trajectories of a CO_2 Gaussian beam with a waist of 3 cm. (With permission from [29]).

orthogonal coordinates (x, y, z) of Fig. 5.9, and we will refer to the plane containing the magnetic axis as the equatorial plane and to the (r, z)-plane (with $r = \sqrt{x^2 + y^2}$) as the poloidal plane.

Since $\theta_s \ll 1$, the two scattering branches of Fig. 5.7 have similar selectivity functions, with maxima near the point $x = 0$, $y = y_0$ (with y_0 the initial y-coordinate of the probing beam). For simplicity, we will consider only the scattering branch corresponding to the $+$ sign in (5.45). We will also refer to the quantity $\varepsilon = (y_0 - r_{ma})/(y_b - r_{ma})$ (with y_b the maximum y-coordinate of the plasma boundary in Fig. 5.9 and r_{ma} the radius of the magnetic axis) as the average normalized radius of the scattering region.

Figure 5.10 shows the instrumental selectivity function along the central ray of the probing beam for $\varepsilon = 0.5$ and three values of k. As expected from (5.43), the width of F is a strongly decreasing function of the fluctuation wave number. The components of $\mathbf{k}$ are displayed

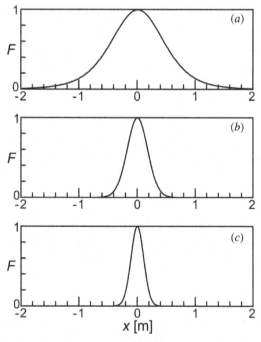

Figure 5.10. Instrumental selectivity for the scattering geometry of Fig. 5.9 with $\varepsilon = 0.5$, $\beta = 14°$ and $k = 2$ (a), $k = 5$ (b) and $k = 8$ (c) cm^{-1}. (With permission from [29]).

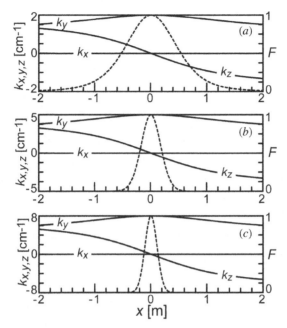

Figure 5.11. Wave vector components of detected fluctuations for the three cases of Fig. 5.10 (dash lines are the instrumental selectivity). (With permission from [29]).

in Fig. 5.11, showing that their spatial profiles are not a sensitive function of k.

As shown in [28], the scattering region is located near the point where the angle β between the wave vector of the probing wave and the magnetic field is minimum, with the width of F an increasing function of β. This is illustrated in Fig. 5.12, showing that the width of F increases with the value of β (which is a growing function of ε as well).

Another way of changing β is by launching the probing beam at an oblique angle with the equatorial plane, as in Fig. 5.13 where the beam is launched perpendicularly to the y-axis as before, but now making an angle (γ) of $\pm 4.5°$ with the x-axis (the $\pm$ sign indicating the up and down direction). To emphasize the role of β, the launching point has been chosen to make the beam symmetric with respect to the equatorial plane. For $\varepsilon = 0.5$, this results in β varying from $9.5°$ for $\gamma = -4.5°$ to $18.5°$ for $\gamma = 4.5°$. The corresponding values of F

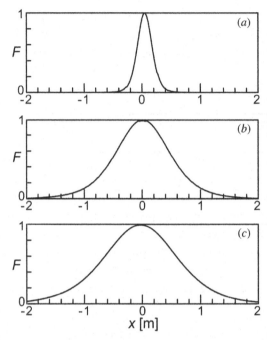

Figure 5.12. Instrumental selectivity for the geometry of Fig. 5.9 with $k = 2\,\mathrm{cm}^{-1}$ and (a) $\varepsilon = 0.15$ ($\beta = 4.3°$), (b) $\varepsilon = 0.5$ ($\beta = 14°$, as in Fig. 5.10), (c) $\varepsilon = 0.7$ ($\beta = 18°$). (With permission from [29]).

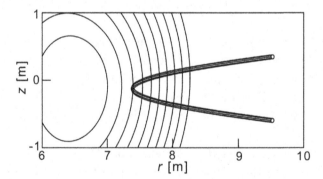

Figure 5.13. As in Fig. 5.9(a) with $\gamma = \pm 4.5°$. The initial point is chosen to make the trajectory symmetric with respect to the equatorial plane. (With permission from [29]).

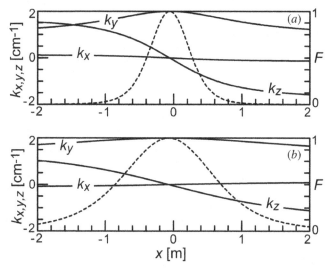

Figure 5.14. Same as in Fig. 5.11 for $k = 2\,\mathrm{cm}^{-1}$, $\varepsilon = 0.15$ and (a) $\gamma = -4.5°\,(\beta = 9.5°)$, (b) $\gamma = 4.5°\,(\beta = 18.5°)$. (With permission from [29]).

for $k = 2\,\mathrm{cm}^{-1}$ are displayed in Fig. 5.14, showing that the width of the instrumental function for $\gamma = 4.5°$ is more than twice that for $\gamma = -4.5°$, proving that the width of F is a decreasing function of β.

So far we have consider the profile of the instrumental function along the probing beam. However, a crucial parameter in fluctuation measurements is the radial resolution. In our case, this can be inferred from the radial footprint (δr) on the equatorial plane of the part of the central ray with $F > 1/e$. This is displayed in Fig. 5.15 showing that δr becomes quickly much smaller than the beam diameter $(2w)$ when $k > 2\,\mathrm{cm}^{-1}$, and thus the radial resolution is essentially the beam diameter.

5.6 Wave Number Resolution

The instrumental selectivity function used so far was defined in Sec. 5.4 as the collection efficiency of waves scattered by fluctuations with the same value of k. In this section, we generalize the definition of selectivity function to include fluctuations with all possible wave numbers. To be more precise, let us consider again scattered waves originating from two points of the probing beam with wave vectors

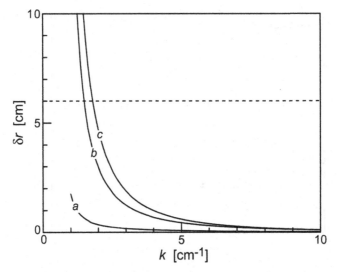

Figure 5.15. Radial footprint (δr) of the part of central ray where $F > 1/e$ for $\gamma = 0°$ and (a) $\varepsilon = 0.15$, (b) $\varepsilon = 0.5$, (c) $\varepsilon = 0.7$. Dash line is the beam diameter ($2w$). (With permission from [29]).

$\mathbf{k}_s^1$ and $\mathbf{k}_s^2$, respectively. From (5.41), we get

$$\frac{\mathbf{k}_s^1 \cdot \mathbf{k}_s^2}{k_i^2} = \cos \alpha$$

$$= \cos\theta_1 \cos\theta_2 + \sin\theta_1 \sin\theta_2 (\cos\varphi_1 \cos\varphi_2 + \sin\varphi_1 \sin\varphi_2),$$
$$(5.47)$$

where θ_1 and θ_2 are the corresponding scattering angles. From this, we obtain

$$\cos\alpha = \cos(\theta_2 - \theta_1) - 2\sin\theta_1 \sin\theta_2 \sin^2(\delta\varphi/2), \qquad (5.48)$$

where $\delta\varphi = \varphi_2 - \varphi_1$. Since both θ_1 and θ_2 are very small, this becomes

$$\alpha^2 \approx (\theta_2 - \theta_1)^2 + 4\theta_2\theta_1 \sin^2(\delta\varphi/2). \qquad (5.49)$$

The mismatch angle between $\mathbf{k}_s^1$ and $\mathbf{k}_s^2$ is made of two terms. The first, as expected, is due to the difference in scattering angles. The second, as before, is due to the spatial variation of the magnetic pitch angle.

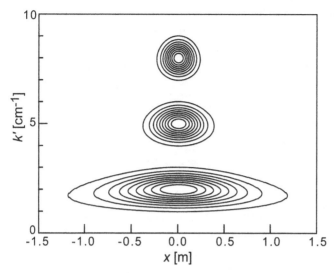

Figure 5.16. Contour plots (nine levels equally spaced from 0.1 to 0.9) of $F(k',x)$ for the three cases of Fig. 5.10 (with $k = 2,5,8\,\text{cm}^{-1}$ from bottom to top). (With permission from [29]).

Following the same procedure that led to (5.43), we obtain a new expression for the instrumental selectivity function

$$F = \exp[-((k' - k)^2 + 4k'k\sin^2(\delta_\varphi/2))/\Delta^2], \qquad (5.50)$$

where $k \approx k_i\theta_1$ is the tuning wave number of the receiver, and $k' \approx k_i\theta_2$ is the wave number of detected fluctuations. For $k' = k$, we recover (5.43).

Contour plots of $F(k',x)$ are displayed in Fig. 5.16 for the same cases of Fig. 5.10, showing that the maximum width of F along the beam trajectory is that given by (5.43) and, as expected, that the wave number resolution is $\approx \pm\Delta$.

The use of probing Gaussian beams implies the use of circular ports with larger radii (r_w) than w. Assuming the conservative criterion $r_w = 2w$, the value of w used so far (3 cm) would require a port much smaller than the size of ITER. Hence for obtaining an improvement in wave number resolution we could envision the use of a larger beam, as in Fig. 5.17 where the contour plots of $F(k',x)$ are displayed for $w = 6\,\text{cm}$.

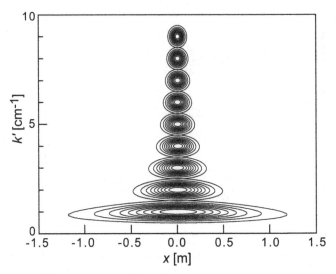

Figure 5.17. Contour plots (nine levels equally spaced from 0.1 to 0.9) of $F(k', x)$ for the scattering geometry of Fig. 5.9 with $\varepsilon = 0.5$, $w = 6\,\mathrm{cm}$ and $k = 1\text{–}9\,\mathrm{cm}^{-1}$ from bottom to top. (With permission from [29]).

In conclusion, scattering of electromagnetic waves has played a major role in the investigation of short-scale turbulence in existing tokamaks. Here we have shown that the use of CO_2 lasers would allow this technique to continue being an important diagnostic tool in the next generation of burning plasma experiments.

Bibliography

[1] Coppi, B. and Rewoldt, in *Advances in Plasma Physics*, Vol. 6, Edited by Simon, A. and Thompson, W. B., Wiley, New York, 1975, p. 421.
[2] Connor, J. W. and Wilson H. R., *Plasma Phys. Control. Fusion* **36**, 719 (1994).
[3] Horton, W., *Turbulent Transport in Magnetized Plasmas*, World Scientific, London, 2012.
[4] Mazzucato, E., *Bull. Am. Phys. Soc.* **20**, 1241 (1975); Princeton University Plasma Physics Laboratory Report MATT-1151 (1975).
[5] Mazzucato, E., *Phys. Rev. Lett.* **36**, 792 (1976).
[6] Tatarski, V. I., *Wave Propagation in a Turbulent Medium*, Dover, New York, 1961.
[7] Mazzucato, E., *Il Nuovo Cimento* **42**, 257 (1966).
[8] Born, M. and Wolf, E., *Principles of Optics*, Pergamon Press, Oxford, 1965.
[9] Lee, W. Y., *Statistical Theory of Communications*, Wiley, New York, 1960.
[10] Mazzucato, E., *Phys. Rev. Lett.* **48**, 1828 (1982).

[11] Mazzucato, E., Smith, D. R., Bell, R. E., Kaye, S. M. *et al.*, *Phys. Rev. Lett.* **101**, 075001 (2008).

[12] Mazzucato, E., Bell, R. E., Ethier, S., Hosea, J. C *et al.*, *Nucl. Fusion* **49**, 055001 (2009).

[13] Brower, D. L., Peebles, W. A., Luhmann, Jr., N. C. and Savage, R. L. *Phys. Rev. Lett.*, **54**, 689 (1985).

[14] Brower, D. L., Peebles, W. A., Kim, S. K., Luhmann, N. C. *et al.*, *Phys. Rev. Lett.* **59**, 48 (1987).

[15] Brower, D. L., Redi, M. H., Tang, W. M., Bravenec, R. V. *et al.*, *Nucl. Fusion* **29**, 1247 (1989).

[16] Crowly, T. and Mazzucato, E., *Nucl. Fusion* **25**, 507 (1985).

[17] Rhodes, T. L., Peebles, W. A., Van Zeeland, M. A., deGrassie, J. S. *et al.*, *Nucl. Fusion* **47**, 936 (2007a).

[18] Smith, D. R., Kaye, S. M., Lee, W., Mazzucato, E. *et al.*, *Phys. Rev. Lett.* **102**, 115002 (2009).

[19] Rhodes, T. L., Peebles, W. A., Van Zeeland, M. A., deGrassie, J. S. *et al.*, *Phys. Plasmas* **14**, 056117 (2007).

[20] Ren, Y., Kaye, S. M., Mazzucato, E., Bell, R. E. *et al.*, *Phys. Rev. Lett.* **106**, 165005 (2011).

[21] Surko, C. M. and Slusher, R. E., *Phys. Rev. Lett.* **37**, 1747 (1976).

[22] Surko, C. M. and Slusher, R. E., *Phys. Fluids* **23**, 2425 (1980).

[23] Truc, A., Quéméneur, A., Hennequin, P., Grésillon, D. *et al.*, *Rev. Sci. Instrum.* **63**, 3716 (1992).

[24] Devynck, P., Garbet, X., Laviron, C., Payan, J. *et al.*, *Plasma Phys. Control. Fusion* **35**, 63 (1993).

[25] Coda, S., Porkolab, M. and Burrell, K. H., *Phys. Lett. A* **273**, 125 (2000).

[26] Coda, S., Porkolab, M. and Burrell, K. H., *Nucl. Fusion* **41**, 1885 (2001).

[27] Hennequin, P., Sabot, R., Honoré, C., Hoang, G. T. *et al.*, *Plasma Phys. Control. Fusion* **46**, B121 (2004).

[28] Mazzucato, E., *Phys. Plasmas* **10**, 753 (2003).

[29] Mazzucato, E., *Plasma Phys. Control. Fusion* **48**, 1749 (2006).

[30] ITER Technical Basis, *ITER EDA Documentation Series No. 24*, IAEA, Vienna, 2002.

CHAPTER 6

NON-COLLECTIVE SCATTERING

The scattering cross-section that we derived in the previous chapter was the result of the hidden assumption that the plasma reacted to the probing wave as a fluid rather than a collection of random particles. In other words, we assumed that the scattered fields of individual electrons were correlated, and thus the average of the square of their vector sum was much larger than the average sum of their square. As a result we got a cross-section scaling like the square of the amplitude of electrons fluctuations. This is why this type of scattering goes under the name of *collective scattering* of electromagnetic waves. In this chapter we will deal with the opposite case — *non-collective scattering* — where the probing wave samples the electrons on a scale length where they appear free. From plasma theory we know that this would occur when the wavelength of fluctuations we try to detect is smaller than the electron Debye length $\lambda_{\mathrm{De}} = \sqrt{\kappa T_e / 4\pi n e^2}$, since no collective oscillations can exist with wavelengths appreciably shorter than λ_{De} [1], i.e., when

$$k\lambda_{\mathrm{De}} \gg 1, \qquad (6.1)$$

where k is the scattering wave number. The reason for doing non-collective scattering of electromagnetic waves is that it provides a way of measuring the density and temperature of electrons — two of the most important plasma parameters. This technique, that in the plasma literature goes under the name of *Thomson Scattering*, was used for the first time in a magnetically confined toroidal plasma on the Model C-Stellarator [2] and soon after on the T-3 Tokamak [3].

The latter experiment represented a turning point in the development
of tokamaks since it established its magnetic confinement scheme as
the front-runner in fusion research.

6.1 Radiation by a Moving Electron

Let us consider the plane electromagnetic wave

$$\mathbf{E}_i = \mathbf{E}_0 \exp[i(\mathbf{k}_i \cdot \mathbf{r} - \omega_i t)], \quad \mathbf{B}_i = \frac{c}{\omega} \mathbf{k}_i \times \mathbf{E}_i, \qquad (6.2)$$

impinging on an electron with the relativistic equation of motion

$$\frac{d}{dt} \frac{m_e \mathbf{v}}{(1 - v^2/c^2)^{1/2}} = -e \left(\mathbf{E}_i + \frac{\mathbf{v} \times \mathbf{B}_i}{c} \right)$$

that can be written as

$$\gamma m_e \dot{\mathbf{v}} + \gamma^3 m_e \mathbf{v} \frac{\mathbf{v} \cdot \dot{\mathbf{v}}}{c^2} = -e \left(\mathbf{E}_i + \frac{\mathbf{v} \times \mathbf{B}_i}{c} \right), \qquad (6.3)$$

where m_e is the electron rest mass, $\gamma = (1 - v^2/c^2)^{-1/2}$ and the over-
dot indicates the time derivative. Taking the scalar product with $\mathbf{v}$
we obtain

$$\mathbf{v} \cdot \dot{\mathbf{v}} = -\frac{e}{\gamma^3 m_e} \mathbf{v} \cdot \mathbf{E}_i$$

that when substituted back into (6.3) yields

$$\dot{\boldsymbol{\beta}} = -\frac{e}{\gamma m_e c} [\mathbf{E}_i - \boldsymbol{\beta}(\boldsymbol{\beta} \cdot \mathbf{E}_i) + \boldsymbol{\beta} \times \mathbf{B}_i], \qquad (6.4)$$

with $\boldsymbol{\beta} = \mathbf{v}/c$.

The radiation from a moving electron — the scattering field —
can be calculated from Maxwell equations expressed in terms of
scalar (ϕ) and vector ($\mathbf{A}$) potentials in the Lorentz gauge [4]

$$\mathbf{B}_s = \nabla \times \mathbf{A},$$

$$\mathbf{E}_s = -\frac{1}{c} \frac{\partial \mathbf{A}}{\partial t} - \nabla \phi, \qquad (6.5)$$

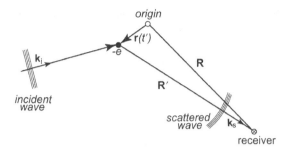

Figure 6.1.   Scattering geometry.

which satisfy the inhomogeneous wave equations

$$\nabla^2 \phi - \frac{1}{c^2}\ddot{\phi} = -4\pi\rho,$$

$$\nabla^2 \mathbf{A} - \frac{1}{c^2}\ddot{\mathbf{A}} = -\frac{4\pi}{c}\mathbf{J}, \tag{6.6}$$

where ρ and $\mathbf{J}$ are the charge and current densities. These equations can be solved using the Green's function [4], which for the scattering geometry of Fig. 6.1 is given by

$$G(R, t | R', t') = \frac{\delta(t' - t + |\mathbf{R}'|/c}{|\mathbf{R}'|}, \tag{6.7}$$

where $\delta(\mathbf{x})$ is the Dirac delta function. This gives

$$\phi = \int \frac{[\rho]_{ret}}{R'} d\mathbf{r},$$

$$\mathbf{A} = \frac{1}{c} \int \frac{[\mathbf{J}]_{ret}}{R'} d\mathbf{r}, \tag{6.8}$$

with the subscript *ret* indicating that the quantity must be evaluated at the retarded time $t' = t - R'/c$.

Charge and current densities for a single electron can be written as

$$\rho(\mathbf{r}, t') = -e\delta(\mathbf{r} - \mathbf{r}(t')), \quad \mathbf{j}(\mathbf{r}, t') = -e\mathbf{v}(t')\delta(\mathbf{r} - \mathbf{r}(t')), \tag{6.9}$$

which when substituted in (6.8) yield the Lienard–Wiechert potentials [4]

$$\phi = -e \left[\frac{1}{(1 - \hat{\mathbf{s}} \cdot \boldsymbol{\beta})^3 R'} \right]_{ret}, \quad \mathbf{A} = -\frac{e}{c} \left[\frac{\mathbf{v}}{(1 - \hat{\mathbf{s}} \cdot \boldsymbol{\beta})^3 R'} \right]_{ret}, \quad (6.10)$$

where $\hat{\mathbf{s}} = \mathbf{R}'/R'$. Finally, by inserting (6.10) into (6.5) we obtain

$$\mathbf{E}_s = -e \left[\frac{(\hat{\mathbf{s}} - \boldsymbol{\beta})(1 - \beta^2)}{(1 - \hat{\mathbf{s}} \cdot \boldsymbol{\beta})^3 R'^2} \right]_{ret} - \frac{e}{c} \left[\frac{\hat{\mathbf{s}} \times [(\hat{\mathbf{s}} - \boldsymbol{\beta}) \times \dot{\boldsymbol{\beta}}]}{(1 - \hat{\mathbf{s}} \cdot \boldsymbol{\beta})^3 R'} \right]_{ret}, \quad (6.11)$$

$$\mathbf{B}_s = \hat{\mathbf{s}} \times \mathbf{E}_s,$$

where for simplicity we have assumed a refractive index (kc/ω) of one.

In scattering measurements, the distance of the detector from the observation plasma region is usually much larger than the characteristic length of the scattering volume. Thus we can neglect the first term of (6.11) $(\propto 1/R'^2)$ compared to the second $(\propto 1/R')$. We can also make the approximation $R' \approx R$ in the denominator of (6.11) and approximate the retarded time to

$$t' = t - \frac{|\mathbf{R} - \mathbf{r}(t')|}{c} \approx t - \frac{R}{c} + \frac{\hat{\mathbf{s}} \cdot \mathbf{r}(t')}{c}, \quad (6.12)$$

where $\hat{\mathbf{s}}$ can now be assumed to be constant in time.

As explained in the introduction, the non-collective scattering of electromagnetic waves is obtained by first getting the contribution of a single electron from (6.4) and (6.11) and then by integrating over the electron distribution.

From (6.2), (6.4) and (6.11), we have

$$\mathbf{E}_s(R, t) = \frac{r_e}{R} \frac{(1 - \beta^2)^{1/2}}{(1 - \hat{\mathbf{s}} \cdot \boldsymbol{\beta})^3} \{ \hat{\mathbf{s}} \times [(\hat{\mathbf{s}} - \boldsymbol{\beta}) \times (\hat{\mathbf{e}} + \boldsymbol{\beta} \times (\hat{\mathbf{i}} \times \hat{\mathbf{e}}) - \boldsymbol{\beta}(\boldsymbol{\beta} \cdot \hat{\mathbf{e}}))] \}$$

$$\times E_0 \exp[i(\mathbf{k}_i \cdot \mathbf{r}(t') - \omega_i t')], \quad (6.13)$$

where $\hat{\mathbf{i}} = \mathbf{k}_i/k_i$, $\hat{\mathbf{e}} = \mathbf{E}_0/E_0$ and $r_e = e^2/m_e c^2$ is the classical electron radius [5, 6]. By expanding the vector cross products, this expression can be written as

$$\mathbf{E}_s(R,t) = \frac{r_e}{R} \frac{(1-\beta^2)^{1/2}}{(1-\beta_s)^3}$$

$$\times \{\boldsymbol{\beta}(\beta_e(1-(\hat{\mathbf{s}}\cdot\mathbf{i})) - (1-\beta_i)(\hat{\mathbf{s}}\cdot\hat{\mathbf{e}})) - \hat{\mathbf{i}}\beta_e(1-\beta_s)$$

$$+\hat{\mathbf{s}}((1-\beta_i)(\hat{\mathbf{s}}\cdot\hat{\mathbf{e}}) + \beta_e((\hat{\mathbf{s}}\cdot\mathbf{i})-\beta_s)) - \hat{\mathbf{e}}(1-\beta_i)(1-\beta_s)\}$$

$$\times E_0 \exp[i(\mathbf{k}_i\cdot\mathbf{r}(t') - \omega_i t')], \tag{6.14}$$

where all quantities are at the retarded time and $\beta_i = \boldsymbol{\beta}\cdot\hat{\mathbf{i}}, \beta_s = \boldsymbol{\beta}\cdot\hat{\mathbf{s}}, \beta_e = \boldsymbol{\beta}\cdot\hat{\mathbf{e}}$. It can be further simplified by using the geometry that is often employed in scattering measurements, where the incident wave has its electric field ($\mathbf{E}_0$) perpendicular to the scattering plane, so that $\hat{\mathbf{s}}\cdot\hat{\mathbf{e}} = \hat{\mathbf{i}}\cdot\hat{\mathbf{e}} = 0$, and a polarizer is used to measure the component of $\mathbf{E}_s$ parallel to $\mathbf{E}_0$. This reduces (6.14) to

$$\mathbf{e}\cdot\mathbf{E}_s(R,t) = \frac{r_e}{R} \frac{(1-\beta^2)^{1/2}}{(1-\beta_s)^3}$$

$$\times \{\beta_e^2(1-\cos\theta_s) - (1-\beta_i)(1-\beta_s)\}$$

$$\times E_0 \exp[i(\mathbf{k}_i\cdot\mathbf{r}(t') - \omega_i t')], \tag{6.15}$$

with $\hat{\mathbf{s}}\cdot\hat{\mathbf{i}} = \cos\theta_s$ and θ_s the scattering angle.

6.2 Scattered Power

The results of the previous section can be written as

$$\mathbf{E}_s(R,t) = \frac{r_e}{R}\mathbf{V}E_i, \tag{6.16}$$

where $\mathbf{V}$ is the vector function from (6.14) or (6.15), with the frequency spectrum of the scattered field given by

$$\mathbf{E}_s(R,\omega_s) = \frac{r_e}{R}\int \mathbf{V}E_i \exp(i\omega_s t)dt. \tag{6.17}$$

Changing the integral to retarded time using $dt = (1-\beta_s)dt'$ (from (6.12)), we get

$$\mathbf{E}_s(R,\omega_s) = \frac{r_e}{R}\int \mathbf{V}E_i \exp[i\omega_s(t' + (R - \hat{\mathbf{s}}\cdot\mathbf{r}(t'))/c](1-\beta_s)dt' \tag{6.18}$$

that may be written as

$$\mathbf{E}_s(R,\omega_s) = E_0 \frac{r_e}{R} \exp(ik_s R)$$

$$\times \int \mathbf{V} \exp[i(\omega t' - \mathbf{k} \cdot \mathbf{r}(t'))](1 - \beta_s) dt', \quad (6.19)$$

where $\omega = \omega_s - \omega_i$, $\mathbf{k} = \mathbf{k}_s - \mathbf{k}_i$ and $\mathbf{k}_s = \hat{s}\omega_s/c$. For constant electron velocity, the time integration of (6.19) yields apart from a constant phase

$$\mathbf{E}_s(R,\omega_s) = E_0 \frac{r_e}{R} \exp(ik_s R) 2\pi (1 - \beta_s) \mathbf{V} \delta(\mathbf{k} \cdot \mathbf{v} - \omega). \quad (6.20)$$

The scattered field may then be written as

$$\mathbf{E}_s(R,t) = E_0 \frac{r_e}{R} \exp(ik_s R)$$

$$\times \int_{-\infty}^{+\infty} (1 - \beta_s) \mathbf{V} \delta(\mathbf{k} \cdot \mathbf{v} - \omega) \exp(-\omega_s t) d\omega_s. \quad (6.21)$$

From a well-known property of delta functions [7] we get

$$\delta(\mathbf{k} \cdot \mathbf{v} - \omega) = \delta(\omega_i(1 - \beta_i) - \omega_s(1 - \beta_s)) = \delta(\omega_r - \omega_s)/(1 - \beta_s), \quad (6.22)$$

where

$$\omega_r = \omega_i \frac{1 - \beta_i}{1 - \beta_s}. \quad (6.23)$$

The frequency integration in (6.21) can then be easily performed to get

$$\mathbf{E}_s(R,t) = E_0 \frac{r_e}{R} \mathbf{V} \exp[i(k_s R - \omega_r t)], \quad (6.24)$$

showing that the frequency of the scattered field seen by the observer is ω_r. This is the result of the Doppler shift of the input wave frequency from the motion of the electron toward the incident wave ($\omega' = \omega_i - \mathbf{k}_i \cdot \mathbf{v}$), together with the additional Doppler shift of the scattered wave from the electron motion toward the observation point.

In fact, from (6.12), taking $\mathbf{r}(t') = \mathbf{r}(0) + \mathbf{v}t'$, we get

$$t' = \frac{t - R/c + \mathbf{s} \cdot \mathbf{r}(0)/c}{1 - \mathbf{s} \cdot \mathbf{v}/c}$$

giving

$$\Delta t' = \frac{\Delta t}{1 - \beta_s},$$

from which we get that the frequency seen by the observer is

$$\frac{\omega'}{1 - \beta_s} = \omega_i \frac{1 - \beta_i}{1 - \beta_s} = \omega_r.$$

From (6.24), we obtain the time-average scattered power per unit solid angle (Ω_s) and unit frequency

$$\frac{d^2 P_s}{d\Omega_s d\omega_s} = \sigma_0 |\mathbf{V}|^2 P_i \delta(\omega_s - \omega_r), \qquad (6.25)$$

where, as in Chapter 5, $\sigma_0 = (e^2/m_e c^2)^2$ and P_i is the mean incident flux. However, since this is the energy per unit time at the receiver while we want the scattered energy per unit time at the electron, we must multiply the right-hand side of (6.25) by the ratio $dt/dt' = 1 - \beta_s$.

Equation (6.25) yields the total scattered power from a group of N electrons with velocity distribution $f(\mathbf{v})$

$$\frac{d^2 P_s}{d\Omega_s d\omega_s} = N \sigma_0 P_i \int |\mathbf{V}|^2 (1 - \beta_s)^2 f(\mathbf{v}) \delta(\mathbf{k} \cdot \mathbf{v} - \omega) d\mathbf{v}, \qquad (6.26)$$

where we have made again use of (6.22).

Using the simplified expression of $\mathbf{V}$ (6.15), from (6.26) we finally obtain

$$\frac{d^2 P_s}{d\Omega_s d\omega_s} = N \sigma_0 P_i \int \left[\frac{1 - \beta_i}{1 - \beta_s}\right]^2 \left[1 - \frac{\beta_e^2(1 - \cos\theta_s)}{(1 - \beta_i)(1 - \beta_s)}\right]^2$$
$$\times (1 - \beta^2) f(\mathbf{v}) \delta(\mathbf{k} \cdot \mathbf{v} - \omega) d\mathbf{v}. \qquad (6.27)$$

The first term in the integrand of (6.27) is $(\omega_s/\omega_i)^2$ and therefore can be taken outside of the integral, while the second term arises from the relativistic depolarization of scattered waves. The scattered power can then be cast in the form

$$\frac{d^2 P_s}{d\Omega_s d\omega_s} = N\sigma_0 P_i S(\omega_s, \theta_s), \qquad (6.28)$$

where we have introduced the spectral density function of scattered radiation

$$S(\omega_s, \theta_s) = \frac{\omega_s^2}{\omega_i^2} \int \left[1 - \frac{\beta_e^2(1 - \cos\theta_s)}{(1 - \beta_i)(1 - \beta_s)} \right]^2$$
$$\times [1 - \beta^2] f(\mathbf{v}) \delta(\mathbf{k} \cdot \mathbf{v} - \omega) d\mathbf{v}. \qquad (6.29)$$

6.3 Spectral Density to Second Order in β

For high temperature plasmas, we must use the relativistic form of the Maxwell distribution function [8]

$$f(\beta) = \frac{\alpha \exp[-2\alpha/(1 - \beta^2)^{1/2}]}{2\pi(1 - \beta^2)^{5/2} K_2(2\alpha)}, \qquad (6.30)$$

where $\alpha = m_e c^2/2\kappa T_e$, K_2 is the modified Bessel function of the second order [9] and κ is the Boltzmann constant. This together with the complicated dependence of the integrand of (6.29) on β makes very difficult to derive a complete analytic expression for the scattered power. One must then resort to some simplifying approximations or to numerical calculations. Examples of the latter are in [10], showing that the main effect of relativistic corrections is a blue shift of the spectrum of scattered waves. This is explained by the fact that while in the frame of a relativistic electron the emitted radiation has a symmetrical donut shape, in the frame of a stationary observer the radiation is preferentially emitted in the forward direction. This together with the upward Doppler frequency shift of waves from electrons moving towards the observer make the spectrum of measured radiation to shift towards the blue side.

It is not difficult to calculate the scattered radiation to second order in β. For this, we begin by taking advantage of the large value of α in present and future fusion experiments for approximating K_2 with [9]

$$K_2(2\alpha) \approx \sqrt{\frac{\pi}{4\alpha}} \exp(-2\alpha) \left(1 + \frac{15}{16\alpha}\right). \qquad (6.31)$$

Also to the second order in β we get

$$\exp[-2\alpha(1-\beta^2)^{-1/2}] \approx \exp(-2\alpha) \exp(-\alpha\beta^2) \left(1 - \frac{3}{4}\alpha\beta^4\right),$$
$$(1-\beta^2)^{-5/2} \approx 1 + \frac{5}{2}\beta^2, \qquad (6.32)$$

so that (6.30) is reduced to

$$f(\beta) \approx \left(\frac{\alpha}{\pi}\right)^{3/2} \exp(-\alpha\beta^2) \left(1 + \frac{5}{2}\beta^2 - \frac{15}{16\alpha} - \frac{3}{4}\alpha\beta^4\right), \qquad (6.33)$$

where $(\alpha/\pi)^{3/2} \exp(-\alpha\beta^2)$ is the non-relativistic Maxwellian distribution function. To second order in β we get

$$(1-\beta^2) \left[1 - \frac{\beta_e^2(1-\cos\theta_s)}{(1-\beta_i)(1-\beta_s)}\right]^2 f(\beta)$$
$$\approx \left(\frac{\alpha}{\pi}\right)^{3/2} \exp(-\alpha\beta^2) \left[1 + \frac{3}{2}\beta^2 - \frac{15}{16\alpha} - \frac{3}{4}\alpha\beta^4 - 2\beta_e^2(1-\cos\theta_s)\right]. \qquad (6.34)$$

Following [10], we introduce the system of orthogonal coordinates $(\beta_e, \beta_\perp, \beta_k)$ of Fig. 6.2, where β_e is perpendicular to the scattering plane and β_k is parallel to the scattering wave vector $\mathbf{k}$, so that $\beta_\perp$ is perpendicular to $\mathbf{k}$. In this coordinate system we have

$$\beta_i = -\beta_k \cos\gamma + \beta_\perp \sin\gamma,$$
$$\beta_s = -\beta_k \cos(\gamma + \theta_s) + \beta_\perp \sin(\gamma + \theta_s), \qquad (6.35)$$
$$\cos\gamma = (\omega_i - \omega_s \cos\theta_s)/kc,$$

where $\pi - \gamma$ is the angle between $\mathbf{k}_i$ and $\mathbf{k}$ as shown in Fig. 6.2, and the last equation comes from $(\mathbf{k}_s - \mathbf{k}_i) \cdot \mathbf{k}_i/kk_i = -\cos\gamma$.

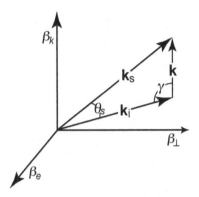

Figure 6.2. β coordinates.

Replacing (6.34) in (6.29) and performing the integration in β_k, we obtain

$$S((\omega/\omega_i), \theta_s, \alpha) = \left(\frac{\alpha}{\pi}\right)^{3/2} \frac{\omega_s^2}{\omega_i^2} \frac{\omega_i}{kc} \int \exp\left[-\alpha\left(\beta_e^2 + \beta_\perp^2 + \frac{\omega^2}{k^2c^2}\right)\right]$$

$$\times \left\{1 - \frac{3}{2}\left(\beta_e^2 + \beta_\perp^2 + \frac{\omega^2}{k^2c^2}\right) - \frac{15}{16\alpha}\right.$$

$$- \frac{3}{4}\alpha\left[\beta_e^4 + \beta_\perp^4 + \frac{\omega^4}{k^4c^4} + 2\left(\beta_e^2\beta_\perp^2 + \beta_e^2\frac{\omega^2}{k^2c^2}\right.\right.$$

$$\left.\left.+ \beta_\perp^2\frac{\omega^2}{k^2c^2}\right)\right] - 2\beta_e^2(1 - \cos\theta_s)\right\} d\beta_e d\beta_\perp,$$

$$(6.36)$$

where we used

$$\delta(\mathbf{k} \cdot \mathbf{v} - \omega) = \delta(kv_k - \omega) = (1/kc)\delta(\beta_k - \omega/kc)$$

and made a change of variable from ω_s to the normalized frequency ω/ω_i.

To proceed, we need to perform integrals of the type

$$\int_{-1}^{+1} \beta^n \exp(-\alpha\beta^2)d\beta,$$

with $n = 0, 2, 4$, which can be expressed in terms of the error function [11]. However, because of the large value of α we can safely extend the integration to $\pm\infty$ and use the well-known integrals of the Gaussian function, so that by performing the last two integration in (6.36) we obtain

$$S((\omega/\omega_i), \theta_s, \alpha) = \left(\frac{\alpha}{\pi}\right)^{1/2} \frac{\omega_s^2}{\omega_i^2} \frac{\omega_i}{kc} \exp\left[-\alpha\frac{\omega^2}{k^2c^2}\right]$$

$$\times \left\{1 - \frac{63}{16}\frac{1}{\alpha} - 3\frac{\omega^2}{k^2c^2} - \frac{3}{4}\alpha\frac{\omega^4}{k^4c^4} - \frac{1}{\alpha}(1 - \cos\theta_s)\right\},$$

$$(6.37)$$

where

$$\frac{\omega_s^2}{\omega_i^2} = \frac{\omega_i^2 + \omega^2 + 2\omega_i\omega}{\omega_i^2} = 1 + 2\frac{\omega}{\omega_i} + \frac{\omega^2}{\omega_i^2},$$

$$\frac{k^2c^2}{\omega_i^2} = \frac{(\omega_i + \omega)^2 + \omega_i^2 - 2\omega_i(\omega_i + \omega)\cos\theta_s}{\omega_i^2} \qquad (6.38)$$

$$= 2\left(1 + \frac{\omega}{\omega_i}\right)(1 - \cos\theta_s) + \frac{\omega^2}{\omega_i^2}.$$

For comparison, the non-relativistic spectral density function is given by [6]

$$S((\omega/\omega_i), \theta_s, \alpha) = \left(\frac{\alpha}{\pi}\right)^{1/2} \frac{1}{\sqrt{2[1 - \cos\theta_s]}} \exp\left[-\frac{\alpha}{2[1 - \cos\theta_s]}\frac{\omega^2}{\omega_i^2}\right].$$

$$(6.39)$$

Figure 6.3 displays the spectral density (6.37) as a function of ω/ω_i for two scattering angle, clearly showing an increased distortion and a blue shift of the spectrum as the electron temperature increases. Thus, for obtaining the temperature of hot plasmas from the best fitting of the measured scattering spectrum one needs to use both sides of the spectral density — not just one side (usually blue) as it is often done in low temperature plasmas.

A comparison with the full numerical calculation of the scattering spectral density indicates that the second-order approximation is quite accurate up to temperatures of 25 keV [10], beyond which higher order of approximation are required. These could be obtained using a similar procedure to that described above, the only difference being a more cumbersome algebra because of the large number of terms that must be retained.

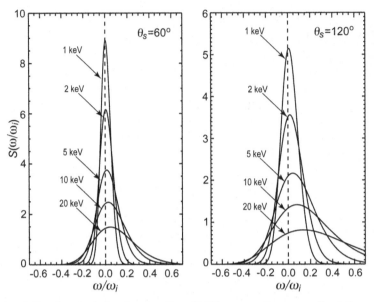

Figure 6.3. Spectral density function (6.37) versus the normalized frequency ω/ω_i for two scattering angle and several values of electron temperature.

6.4 Non-Collective Scattering in Magnetized Plasmas

The results of the previous section stem for the assumption made in (6.20) of a constant electron velocity. This seems to prevent the use of the results of the previous section in magnetized plasmas, where the velocity of charged particles is far from being constant. Fortunately, this assumption does not impair the validity of our results in most scattering configurations, as we will demonstrate in the following.

As in the case without magnetic field, the analysis of non-collective scattering in a magnetized plasma must begin from the equation of motion of a single electron

$$\dot{\boldsymbol{\beta}} = -\frac{e}{\gamma m_e c}[\mathbf{E}_i - \boldsymbol{\beta}(\boldsymbol{\beta} \cdot \mathbf{E}_i) + \boldsymbol{\beta} \times \mathbf{B}], \qquad (6.40)$$

where now $\mathbf{B}$ includes a constant component ($\mathbf{B}_0$) as well as the magnetic field of the incident wave ($\mathbf{B}_i$). We know that the acceleration due to the former causes the emission of cyclotron radiation with frequencies much smaller than the scattering frequencies in which we are interested in this chapter, usually in the visible or in the near infrared. We are therefore interested only in the acceleration from the incident wave so that we can write (6.40) as in (6.4)

$$\dot{\boldsymbol{\beta}} = -\frac{e}{\gamma m_e c}[\mathbf{E}_i - \boldsymbol{\beta}(\boldsymbol{\beta} \cdot \mathbf{E}_i) + \boldsymbol{\beta} \times \mathbf{B}_i], \qquad (6.41)$$

with the only difference that now $\boldsymbol{\beta}$ depends on the unperturbed electron helical motion in the constant magnetic field. The analysis then proceed as before up to

$$\mathbf{E}_s(R, \omega_s) = E_0 \frac{r_e}{R} \exp(ikR)$$

$$\times \int \mathbf{V}(1 - \beta_s) \exp[i(\omega t' - \mathbf{k} \cdot \mathbf{r}(t'))]dt', \quad (6.42)$$

where again the vector $\mathbf{V}$ takes into account the electron helical orbit, which in the system of orthogonal coordinates (x, y, z) with the z-axis along $\mathbf{B}_0$ is given by

$$\begin{aligned}
\mathbf{r}(t') &= \rho_e(\hat{\mathbf{x}} \cos(\omega_c t') + \hat{\mathbf{y}} \sin(\omega_c t') + v_{||} t' \hat{\mathbf{z}}, \\
\boldsymbol{\beta} &= \beta_\perp(-\hat{\mathbf{x}} \sin(\omega_c t') + \hat{\mathbf{y}} \cos(\omega_c t')) + \beta_{||}\hat{\mathbf{z}},
\end{aligned} \qquad (6.43)$$

where $\omega_c = \Omega_c/\gamma$, with $\Omega_c = eB_0/m_e c$, is the electron cyclotron frequency, $\rho_e = v_\perp/\omega_c$ is the electron Larmor radius and $v_{||}$ and $v_\perp$ are the electron velocity parallel and perpendicular to $\mathbf{B}_0$, respectively. In (6.43) we have neglected all terms giving rise to a constant phase of $\mathbf{E}_s$, such as position and phase of the electron orbit at $t' = 0$. Substituting $\mathbf{r}$ in the exponential term of (6.42) yields

$$\exp[i(\omega t' - \mathbf{k} \cdot \mathbf{r}(t'))]$$

$$= \exp\left[i\left([\omega - ck_{||}\beta_{||}]t' - k_\perp \frac{\beta_\perp c}{\omega_c} \sin(\omega_c t')\right)\right], \quad (6.44)$$

where we have chosen the axes so that the y-component of $\mathbf{k}$ is zero. The integrand of (6.42) contains a periodic part that we express in terms of its Fourier components

$$(1 - \beta_s)\mathbf{V} \exp\left[i\left(-k_\perp \frac{\beta_\perp c}{\omega_c} \sin(\omega_c t')\right)\right] = \sum_{n=-\infty}^{\infty} \mathbf{a}_n \exp(-in\omega_c t'),$$

$$(6.45)$$

giving

$$\mathbf{E}_s(R, \omega_s) = E_0 \frac{r_e}{R} \exp(ikR) 2\pi \sum_{n=-\infty}^{\infty} \mathbf{a}_n \delta(\omega - ck_{||}\beta_{||} - n\omega_c). \quad (6.46)$$

The n-harmonic (ω_{sn}) of $\mathbf{E}_s$ is given by

$$(\omega_{sn} - \omega_i) - (\omega_{sn}\hat{\mathbf{s}} - \omega_i\hat{\mathbf{i}}) \cdot \boldsymbol{\beta}_{||} - n\omega_c = 0$$

and thus

$$\omega_{sn} = \frac{\omega_i(1 - \hat{\mathbf{i}} \cdot \boldsymbol{\beta}_{||}) + n\omega_c}{1 - \hat{\mathbf{s}} \cdot \boldsymbol{\beta}_{||}}. \quad (6.47)$$

To see the implications (6.46), let us consider the simple case of non-relativistic scattering where $\mathbf{V} = \hat{\mathbf{s}} \times (\hat{\mathbf{s}} \times \hat{\mathbf{e}})$ so that from the Bessel relation [9]

$$e^{iz\sin(\phi)} = \sum_{n=-\infty}^{\infty} J_n(z)e^{in\phi},$$

where $J_n(z)$ is the Bessel function of the first kind of order n, we get

$$\mathbf{a}_n = \hat{\mathbf{s}} \times (\hat{\mathbf{s}} \times \hat{\mathbf{e}}) J_n(k_\perp \rho_e). \quad (6.48)$$

The scattered power from a single electron averaged over a time longer than $1/\Omega_c$ is

$$\frac{d^2 P_s}{d\Omega_s d\omega_s} = P_i \sigma_0 |\hat{\mathbf{s}} \times (\hat{\mathbf{s}} \times \hat{\mathbf{e}})|^2 \sum_{n=-\infty}^{\infty} J_n^2(k_\perp \rho_e) \delta(\omega_s - \omega_{sn}), \quad (6.49)$$

where again P_i is the mean incident Poynting flux. From this we get the total scattered power from a group of N electrons with Maxwellian velocity distribution $f(\mathbf{v})$

$$\frac{d^2 P_s}{d\Omega_s d\omega_s} = P_i N \sigma_0 |\hat{\mathbf{s}} \times (\hat{\mathbf{s}} \times \hat{\mathbf{e}})|^2$$

$$\times \sum_{n=-\infty}^{\infty} \int J_n^2(k_\perp \rho_e) f(\mathbf{v}) \delta(\omega - k_{||}v_{||} - n\Omega_c) d\mathbf{v}, \quad (6.50)$$

which for the case of a Maxwell velocity becomes

$$
\frac{d^2 P_s}{d\Omega_s d\omega_s} = P_i N \sigma_0 |\hat{\mathbf{s}} \times (\hat{\mathbf{s}} \times \hat{\mathbf{e}})|^2 \sum_{n=-\infty}^{\infty} \frac{1}{\pi^{3/2} v_t^3} \int_{-\infty}^{\infty} \int_0^{\infty} J_n^2(k_\perp \rho_e)
$$
$$
\times \exp\left(-(v_\perp^2 + v_\parallel^2)/v_t^2\right) \delta(\omega - k_\parallel v_\parallel - n\Omega_c) 2\pi v_\perp dv_\parallel dv_\perp,
$$

$$(6.51)$$

where $v_t = (2\kappa T_e/m_e)$. Finally, from the Bessel relation [11]

$$
\int_0^{\infty} J_l^2(bt) \exp(-p^2 t^2) t \, dt = \frac{1}{2p^2} \exp\left[-\left(\frac{b^2}{2p^2}\right)\right] I_l\left(\frac{b^2}{2p^2}\right), \quad (6.52)
$$

where $I_l(z)$ is the modified Bessel function of the first kind of order l, we obtain

$$
\frac{d^2 P_s}{d\Omega_s d(\omega_s/\Omega_c)} = \frac{P_i N \sigma_0}{\pi^{1/2}} |\hat{\mathbf{s}} \times (\hat{\mathbf{s}} \times \hat{\mathbf{e}})|^2 \frac{\Omega_c}{k_\parallel v_t} \exp\left(-\frac{k_\perp^2 v_t^2}{2\Omega_c^2}\right)
$$
$$
\times \sum_{n=-\infty}^{\infty} I_n\left(\frac{k_\perp^2 v_t^2}{2\Omega_c^2}\right) \exp\left[-\frac{(\omega/\Omega_c - n)^2}{(k_\parallel v_t/\Omega_c)^2}\right]. \quad (6.53)
$$

This shows that the spectrum consists in a series of peaks separated from the incident frequency by harmonics of the cyclotron frequency. Each peak has a width $k_\parallel v_t$ and hence can be detected only if

$$
|k_\parallel v_t| < \Omega_c, \quad (6.54)
$$

which imposes a limit to the Doppler broadening caused by the motion of electron along the magnetic field lines. This is indeed a very restrictive condition. For instance, if we take a ruby laser as the source of the incident wave ($\lambda_i = 694.3\,\text{nm}$), a scattering angle of $30°$, a plasma with an electron temperature of $100\,\text{eV}$ and a magnetic field of $50\,\text{kG}$ ($\Omega_c = 9 \times 10^{11}\,\text{rad s}^{-1}$), according to (6.54) the spectrum modulation is visible only if the angle between the scattering wave vector $\mathbf{k}$ and the perpendicular to the magnetic field is less than $1.8°$.

The modulation of the scattered spectrum for different values of $p = |k_\parallel v_t|/\Omega_c$ is illustrated in Fig. 6.4, showing how quickly the modulation fades away as p is raised from 0.1 to 0.5.

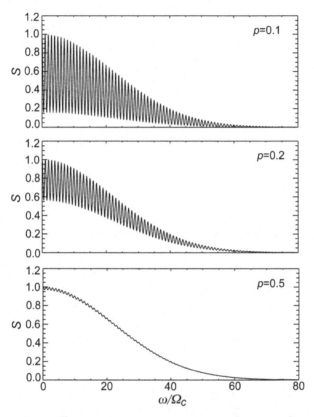

Figure 6.4. Non-collective scattering spectrum as a function of $p = |k_{\parallel} v_t|/\Omega_c$ in a magnetized plasma with an electron temperature of 100 eV and a magnetic field of 50 kG. The probing wave is from a ruby laser and the scattering angle is 30°.

The relativistic case was discussed in [12] where it was found — as expected — that the spread in the values of γ contributes to the broadening of spectral peaks as well. Thus, the visibility condition for the modulation of the scattering spectrum by the magnetic field is even more restrictive than (6.54). Nevertheless, as the case in Fig. 6.4 demonstrates very clearly, the condition (6.54) is sufficient for guaranteeing that the spectrum of scattered waves in thermonuclear plasmas is completely unaffected by the magnetic field.

In conclusion, the modulation of the scattering spectrum by the magnetic field can be detected only in very low temperature plasmas [13]. Thus, the measurements of densities and temperatures of hot magnetized plasmas with non-collective scattering of electromagnetic waves are not affected by the helical orbits of electrons.

6.5 Scattering Measurements

Non-collective scattering of electromagnetic waves is one of the most powerful techniques for the measurement of electron density and temperature in hot plasmas. It is widely used in thermonuclear fusion research in spite of the great technical difficulties that one encounter in its use — all caused by the smallness of the scattering cross-section.

From the previous sections, the order of magnitude of the fraction of scattered power within the solid angle $\Delta\Omega$ is $P_s/P_i \approx \sigma_0 n_e l \Delta\Omega$, where l is the length of the scattering volume. For $n_e = 5 \times 10^{19} \mathrm{m}^{-3}$, $l = 1\,\mathrm{cm}$ and $\Delta\Omega = 10^{-2}\mathrm{sr}$, we get $P_s/P_i \approx 4 \times 10^{-14}$. Such a small scattered power is the source of most of the difficulties in performing non-collective scattering of electromagnetic waves.

The first condition is that the energy delivered by the incident wave in the required measurement resolution time must be large enough to produce a sufficient number of scattered photons to overcome the statistical noise. The second is that the power of the incident wave must be sufficiently high to overcome the background radiation noise from plasma bremsstrahlung and emission from impurities. The conclusion is that what is required is high incident power as well as high energy, which implies the use of short pulse lasers. For instance, for a ruby laser with the energy of $10\,\mathrm{J}$ and a pulse length of $20\,\mathrm{ns}$, the number of scattered photons in the above case is 1.5×10^6. Even using detectors with quantum efficiency as low as 2.5×10^{-2}, this amount of scattered photons would produce an adequate number of photoelectrons for a significant measurement in the presence of the above-mentioned background noise [6].

Figure 6.5 shows the schematic diagram of laser scattering used in early experiments [2, 3], where a ruby laser beam is launch vertically into the plasma and the scattered radiation at $90°$ is collected by an array of lenses and sent to standard polychromators, where the spectrum is measured with a set of phototubes. Because of the small magnitude of the scattering cross-section, particular care must

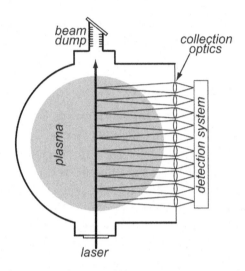

Figure 6.5.   Schematic diagram of laser scattering in early experiments.

be given to the absorption of the unscattered beam that is sent into a beam dump for minimizing the pollution of measured signals.

As already mentioned in the introduction, the scattering results in the T-3 Tokamak [3] represented a milestone in the history of thermonuclear research. The main result of this experiment was to show that the high electron temperature of 1 keV, inferred from other measurements, was indeed a true plasma temperature and not the result of a minority of high-energy electrons riding on the tail of a low temperature distribution.

The scattering scheme of Fig. 6.5 has a variety of problems. The first is that it provides a measurement of the electron temperature only at a single time. The second problem is that it makes use of cumbersome polychromators and phototubes with low quantum efficiency. The third is that it requires a large access port. All of these problems are taken care in modern scattering experiments [14–20]. The first by using rapidly pulsing (10–100 Hz) Nd:YAG lasers ($\lambda = 1064$ nm) that allow scattering measurements many times during a plasma discharge. The second by using compact polychromators utilizing a set of high performance interference filters, with each selecting a frequency band of the scattered radiations and reflecting the rest to another filter. Silicon avalanche photodiodes with very

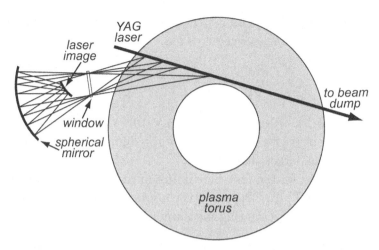

Figure 6.6. Laser scattering geometry with small plasma accessibility. The spherical mirror makes an image of the laser beam over a bundle of fiber optics (not shown) carrying the scattering signals to polychromators.

high quantum efficiency (40–80%) are used for detectors as well. Finally, the plasma accessibility is minimized by using the scattering geometry of Fig. 6.6, where a large spherical mirror collects the scattering radiation through a narrow window and makes an image of the laser beam over a bundle of fiber optics that carry the scattering signals to a set of polychromators.

Finally, the ultimate reduction in plasma accessibility is achieved by using the LIDAR (light detection and ranging) technique, where backscattering from a very short laser pulse is measured with a fast detector. The localization of the measurement is then determined from the time (t) of observation, i.e., the time that the laser takes to travel to the point of observation plus the time for the scattered radiation to travel back to the detector. Assuming for simplicity that the latter is located at the launching point, at the time of observation the laser has traveled the distance $L = ct/2$, from which one gets the measurement location. The spatial resolution depends on the uncertainty (δt) in the value of t, which is a combination of the time duration of the laser pulse (τ_L) and the time response of the detector (τ_D). It can be estimated from

$$\delta t = (\tau_L^2 + \tau_D^2)^{1/2}.$$

If the detector time response is matched to that of the laser pulse, the uncertainty on the measurement of t is $\delta t = \sqrt{2}\tau_L$, and thus the spatial resolution is

$$\delta L = c\tau_L/\sqrt{2}.$$

As an example, $\delta L \approx 10\,\mathrm{cm}$ for $\tau_L = 0.5 \times 10^{-9}\,\mathrm{s}$, which gives an adequate number of resolved points across the diameter of a plasma with a minor radius larger than $1\,\mathrm{m}$.

LIDAR was used for the first time in fusion research on the Joint European Torus (JET) [21, 22], and it is considered to be the leading candidate for Thomson scattering measurements in ITER [23–25].

Bibliography

[1] Bernstein, I. B., Trehan, S. K. and Weenink, M. P. H., *Nucl. Fusion* **4**, 61 (1964).
[2] Dimock, D. and Mazzucato, E., *Phys. Rev. Lett.* **20**, 713 (1968).
[3] Peacock, N. J., Robinson, D. C., Forrest, M. J., Wilcock, P. D. and Sannikov, V. V., *Nature* **224**, 488 (1969).
[4] Jackson, J. D., *Classical Electrodynamics*, Wiley, New York, 1998.
[5] Sheffield, J., *Plasma Phys.* **14**, 783 (1972).
[6] Sheffield, J., *Plasma Scattering of Electromagnetic Radiation*, Academic Press, New York, 1975.
[7] Dirac, P., *Principle of Quantum Mechanics*, Clarendon Press, Oxford, 1958.
[8] Synge, J. L, *The Relativistic Gas*, North-Holland, Amsterdam, 1956.
[9] Abramowitz, M. and Stegun, I. A., *Handbook of Mathematical Functions*, Dover, New York, 1965.
[10] Matoba, T., Itagaki, T., Yamauchi, T. and Funahashi, A., *Japan. J. Appl. Phys.* **18**, 1127 (1979).
[11] Gradshteyn, I. S. and Ryzhik, I. M., *Table of Integrals, Series, and Products*, Academic Press, New York, 2000.
[12] Nee, S. F., Pechacek, R. E. and Trivelpiece, A. W., *Phys. Fluids* **12**, 2651 (1969).
[13] Carolan, P. G. and Evans, D. E., *Plasma Phys.* **13**, 947 (1971).
[14] Rohr, H., Steuer, K. H. and ASDEX Team, *Rev. Sci. Instrum.* **59**, 1875 (1988).
[15] Carlstrom, G. L., DeBoo, J. C., Evanko, R. *et al.*, *Rev. Sci. Instrum.* **61**, 2858 (1990).
[16] Murmann, H., Gotsch, S., Rohr, H. *et al.*, *Rev. Sci. Instrum.* **63**, 4941 (1992).
[17] Carlstrom, T. N., Campbell, G. L., DeBoo, J. C. *et al.*, *Rev. Sci. Instrum.* **63**, 4901 (1992).
[18] Johnson, D., Bretz, N., LeBlanc B. P. *et al.*, *Rev. Sci. Instrum.* **70**, 776 (1999).

[19] LeBlanc, B. P., Bell, R. E., Johnson, D. W. *et al.*, *Rev. Sci. Instrum.* **74**, 1659 (2003).
[20] Scannel, R., Walsh, M. J., Dunstan, M. R. *et al.*, *Rev. Sci. Instrum.* **81**, 10D520 (2010).
[21] Salzmann, H., Bundgaard, J., Gadd, A. *et al.*, *Rev. Sci. Instrum.* **59**, 1451(1985).
[22] Gowers, C., Brown, B. W., Fajemirokun, H. *et al.*, *Rev. Sci. Instrum.* **66**, 471(1995).
[23] Walsh, M. J., Beurskens, M., Carolan, P. J. *et al.*, *Rev. Sci. Instrum.* **77**, 10E525 (2006).
[24] Giudicotti, L., Pasqualotto, R., Afier, A. *et al.*, *Fusion Eng. Design* **86**, 198 (2011).
[25] Naylor, G. A., Scannel, R., Beurskens, M. *et al.*, *J. Instrum.* **7**, C03043 (2012).

CHAPTER 7

PLASMA REFLECTOMETRY

Density measurements play an essential role in the study and operation of magnetically confined plasmas. In existing large devices, such as tokamaks, the canonical tools for the measurement of the electron density are laser multichannel interferometry (Chapter 4) and Thomson scattering (Chapter 6), two of the most reliable methods in thermonuclear fusion research. Unfortunately, these are also two of the most demanding techniques for plasma accessibility, which ironically is a problem that seems to escalate with the plasma size. Indeed, plasma accessibility will become extremely difficult in a fusion reactor, where only the simplest diagnostics will survive.

Microwave reflectometry, an offspring of radar techniques used in ionospheric studies [1, 2], is a method where the plasma density is inferred from the group delay of electromagnetic waves that are reflected by a plasma cutoff. Since the first application of this method to laboratory plasmas [3, 4] and the proposal of employing frequency modulated continuous wave (FMCW) radar techniques in combination with swept millimeter-wave oscillators [5, 6], microwave reflectometry has matured very quickly, and nowadays finds extensive use in tokamak research. Its modest requirement for plasma accessibility and the possibility of conveying microwaves to a remote location make this an ideal method for a fusion reactor.

Another area where microwave reflectometry can play a crucial role is in the detection of short-scale turbulence, which appears to be the cause of anomalous transport in high temperature plasmas. As a matter of fact, the first use of reflectometry in tokamaks was not for the measurement of plasma density, but for the detection of density fluctuations [7]. It provided the first evidence for the existence of a small-scale and broadband turbulence in tokamak plasmas. Reflectometry was then replaced by microwave and laser scattering techniques [8, 9] that were capable of providing a more direct measure of the spectrum of turbulence. Unfortunately, subsequent measurements [10, 11] showed that the scale length of turbulence in tokamaks increases with plasma size, to the point that scattering methods, with their poor spatial resolution when the turbulence scale length is larger than the wavelength of the probing beam, were no longer capable of separating core from edge phenomena. This led to the reemergence of microwave reflectometry as a diagnostic with good resolution for long wavelength fluctuations.

In this chapter, we will review the application of microwave reflectometry to the measurement of density and its fluctuations in magnetically confined plasmas [12].

7.1   Introduction

The cutoffs that are usually employed in plasma reflectometry are the O and R cutoffs of Chapter 2, whose frequencies are

$$
\begin{aligned}
\omega_O &= \omega_p, \\
\omega_R &= (\omega_c^2/4 + \omega_p^2)^{1/2} + \omega_c/2.
\end{aligned}
\tag{7.1}
$$

In tokamaks, for minimizing the size of needed ports, the probing wave is usually launched from an outside position near the equatorial plane in direction quasi-perpendicular to the magnetic surfaces. Since at the plasma edge $\omega_p \approx 0$, the O cutoff has a maximum inside the plasma, and consequently it cannot be used for probing the full density profile when the wave is launched from only one side of the torus. On the contrary, $\omega_R \, (> \omega_c)$ has a finite value at both low and high field sides of the torus, with the latter being the largest. However, since from (7.1) we get

$$
\frac{\partial \omega_R}{\partial r} = \frac{\omega_p^2/l_n - \omega_R \omega_c/r}{2\omega_R - \omega_c},
\tag{7.2}
$$

where $l_n = n_e/(\partial n_e/\partial r)$ is the density scale length, $\partial \omega_R/\partial r = 0$ for $\omega_p^2 = \omega_R \omega_c l_n/r$, which in tokamak plasmas occurs only for densities exceeding the density limit (Sec. 1.4). Hence, the full density profile of tokamak plasmas can in principle be probed by launching extraordinary waves from the low-field side of the torus.

To avoid strong absorption, the probing wave cannot cross a plasma region where its frequency is equal to one of the first harmonics of ω_c. However, this is not sufficient because of kinetic effects on the cyclotron resonance condition [13, 14]

$$\omega - n\omega_c/\gamma - k_{||}v_{||} = 0, \tag{7.3}$$

where $k_{||}$ and $v_{||}$ are respectively, the wave number and electron velocity components along the magnetic field, and γ is the relativistic gamma factor. This resonance condition imposes to the Doppler shifted wave frequency to be equal to one of the relativistic harmonics of the electron cyclotron frequency. It can be written in dimensionless form as

$$\gamma - n\omega_c/\omega - N_{||}u_{||} = 0, \tag{7.4}$$

where $N_{||} = k_{||} c/\omega$ and $u_{||} = p_{||}/m_e c$, with $p_{||}$ the component of the momentum along the magnetic field and n is the harmonic number. This shows that in high temperature plasmas the cyclotron resonance is down shifted by the relativistic change of the electron mass even in the case of quasi-perpendicular propagation ($N_{||} \approx 0$) and thus wave absorption may occur for $\omega \neq n\omega_c$. The theory of electron cyclotron waves will be reviewed in the next chapter. Here, it is sufficient to mention how reflectometry is affected by wave absorption.

The first result is that the Hermitian part of the dielectric tensor of hot plasmas is different from the dielectric tensor of cold plasmas, and thus the cutoff frequencies (7.1) are modified. In [15], it was shown that for temperatures of interest to fusion research, this phenomenon can be reproduced using the cold dielectric tensor (2.39) together with a simple change of the classical electron mass (m_e)

$$m = m_e \left(1 + \frac{5}{\mu}\right)^{1/2}, \tag{7.5}$$

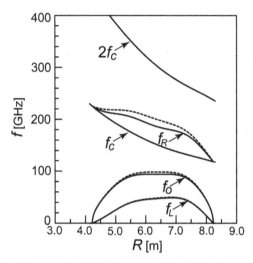

Figure 7.1.   Cyclotron and cutoff frequencies for the plasma of Fig. 7.2. Solid and dash lines are from the cold and hot plasma theory, respectively.

where $\mu = m_e c^2 / T_e$. This affects mostly the R cutoff, as demonstrated by Fig. 7.1 showing the cutoff frequencies of a plasma with the density and temperature profiles of Fig. 7.2.

A second result of the theory of electron cyclotron waves in hot plasmas is the introduction of an anti-Hermitian component of the dielectric tensor, which results in a complex wave vector $\mathbf{k} = \mathbf{k}_r + i\mathbf{k}_i$. In Sec. 3.2, we have seen that for the case of reflectometry, where $|\mathbf{k}_r| \gg |\mathbf{k}_i|$, it is possible to reformulate the theory of geometrical optics with new ray equations that are expressed in terms of the real part of the dispersion relation together with a wave damping given by the integral $\int \mathbf{k}_i \cdot d\mathbf{s}$ along the ray trajectory.

Figure 7.3 shows the ray trajectory of a probing wave with frequency of 185 GHz in the plasma of Fig. 7.2. The wave is reflected by the R cutoff (solid line) that, as the figure clearly illustrates, is drastically modified by relativistic effects. The total wave absorption is displayed in Fig. 7.4 as a function of the reflection point, showing a constant increase as the radius of reflection decreases (i.e., as the wave frequency increases). However, most of the absorption does not occur near the reflection point, where it is very small, but rather near the temperature maximum where the value of γ is large. This

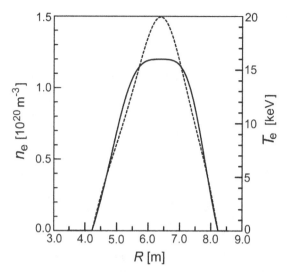

Figure 7.2. Electron density (solid) and temperature (dash) of a tokamak with a toroidal magnetic field of 5.3 T.

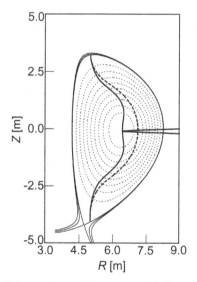

Figure 7.3. Ray trajectory of a probing wave with frequency of 185 GHz in the tokamak plasma of Fig. 7.2. Wave reflection is from the R cutoff (dash line is the cutoff of cold plasma theory).

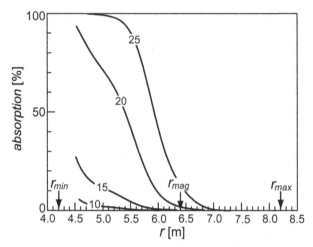

Figure 7.4. Total wave absorption as a function of reflecting radius for the *R* cutoff. Labels are the values of central temperatures; r_{min} and r_{max} are, respectively, the minimum and maximum radii of the plasma boundary and r_{mag} is the radius of the magnetic axis.

is because, as the frequency of the probing wave increases, the value of $2\omega_c/\omega$ near the plasma center decreases, which causes an increase in the absorption of the probing wave. The conclusion that we draw from Fig. 7.4 is that high temperatures preclude the possibility of probing the full density profile even for the *R* cutoff.

In the case of the *O* cutoff, wave absorption remains relatively small at the temperatures of fusion plasmas. However, this is simply due to the fact that the reflection point cannot go beyond the point where the density — and hence the temperature — has its maximum.

In conclusion, neither the *R* nor the *O* cutoff can be used for probing the full density profile in hot fusion plasmas.

7.2 Density Measurements

In microwave reflectometry, the basic information for inferring the plasma density is the phase delay of a probing wave after its reflection from a plasma cutoff. As already mentioned, in general the wave is launched along a trajectory quasi-parallel to the plasma density gradient, and thus quasi-perpendicular to the magnetic field. These conditions are not strictly necessary, but they greatly simplify the collection and analysis of data. It is assumed that the phase delay is given by the approximation of geometrical optics

$$\phi(\omega) = 2 \int_{x_0}^{x_c(\omega)} k(\omega, x)\, dx + \phi_0(\omega)$$

$$= 2\frac{\omega}{c} \int_{x_0}^{x_c(\omega)} N(\omega, x)\, dx + \phi_0(\omega), \qquad (7.6)$$

where x_0 and $x_c\,(> x_0)$ are, respectively, the coordinates of the plasma boundary and the cutoff layer, and $\phi_0(\omega)$ is the phase delay that the wave suffers outside of the plasma. The latter can be either calculated or measured independently. To simplify the notations, in the following we will assume that $\phi_0(\omega)$ has been subtracted from the measured phase, and we will denote by $\phi(\omega)$ the first term on the right-hand side of (7.6).

Since the approximation of geometrical optics is not valid near a cutoff, we need to justify the use of (7.6). For this, let us consider a plane-stratified plasma, which is a good approximation for the case under discussion since the scale-length of fusion plasmas is much larger than the dimensions of the probing beam — meters vs. centimeters. We then assume that in the system of orthogonal coordinates (x, y, z), the probing wave propagates in the x-direction perpendicularly to the plasma strata with the magnetic field in the z-direction. From Maxwell equations, one can easily obtain the equation for the wave electric field $\mathbf{E}(x)$

$$\nabla^2 \mathbf{E} - \nabla(\nabla\!\cdot\!\mathbf{E}) + (\omega^2/c^2)\boldsymbol{\varepsilon} \cdot \mathbf{E} = 0, \qquad (7.7)$$

where $\boldsymbol{\varepsilon}$ is the dielectric tensor (2.39), giving

$$\frac{\omega^2}{c^2}(\varepsilon_{xx} E_x + \varepsilon_{xy} E_y) = 0,$$

$$\frac{d^2 E_y}{dx^2} + \frac{\omega^2}{c^2}(\varepsilon_{xx} E_y - \varepsilon_{xy} E_x) = 0, \qquad (7.8)$$

$$\frac{d^2 E_z}{dx^2} + \frac{\omega^2}{c^2}\varepsilon_{zz} E_z = 0.$$

The combination of the first two and the third of these equations can be written as follows:

$$\frac{d^2 E_y}{dx^2} + \frac{\omega^2}{c^2} N_X^2 E_y = 0,$$

$$\frac{d^2 E_z}{dx^2} + \frac{\omega^2}{c^2} N_O^2 E_z = 0,$$

(7.9)

where N_O and N_X are the refractive indexes (2.43) for the ordinary and extraordinary modes, respectively. Thus, the two modes have a similar equation that for simplicity we write as

$$\frac{d^2 E}{dx^2} + k_0^2 \varepsilon(x) E = 0,$$

(7.10)

with $k_0 = \omega/c$ and ε the wave permittivity (the square of the refractive index N).

We consider first the case of a linear permittivity $\varepsilon(x) = 1 - x/x_c$, for which (7.6) gives $\phi = 4 k_0 x_c/3$. With the change of variables,

$$\zeta = \left(\frac{k_0^2}{x_c}\right)^{1/3} (x - x_c) = -\left(\frac{k_0}{|d\varepsilon/dx|}\right)^{2/3} \varepsilon(x),$$

(7.11)

(7.10) becomes

$$\frac{d^2 E}{d\zeta^2} - \zeta E = 0.$$

(7.12)

Two independent solutions of this equation are the Airy functions $Ai(\zeta)$ and $Bi(\zeta)$ (Fig. 7.5), with asymptotic limits [16]

$$Ai(\zeta) \approx \frac{1}{\pi^{1/2}(-\zeta)^{1/4}} \sin\left(\frac{2}{3}(-\zeta)^{3/2} + \frac{\pi}{4}\right),$$

$$Bi(\zeta) \approx \frac{1}{\pi^{1/2}(-\zeta^{1/4})} \cos\left(\frac{2}{3}(-\zeta)^{3/2} + \frac{\pi}{4}\right),$$

(7.13)

for $\zeta \ll 0$, and

$$Ai(\zeta) \approx \frac{1}{2\pi^{1/2}\zeta^{1/4}} \exp\left(-\frac{2}{3}\zeta^{3/2}\right),$$

$$Bi(\zeta) \approx \frac{1}{2\pi^{1/2}\zeta^{1/4}} \exp\left(\frac{2}{3}\zeta^{3/2}\right),$$

(7.14)

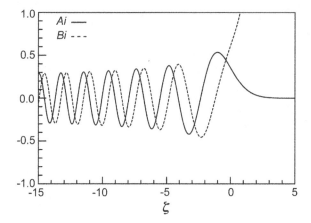

Figure 7.5. Plot of $Ai(\zeta)$ (solid line) and $Bi(\zeta)$ (dash line).

for $\zeta \gg 0$. The accuracy of these limits is better than 1% for $|\zeta| \geq 5$, thus their validity at the plasma edge ($x = 0$) is assured for $k_0 \geq 5^{3/2}/x_c$, a condition that is easily satisfied in plasmas of interest for fusion research.

The general solution of (7.12) is a linear combination of the Airy functions. However, since $Bi(\zeta) \to \infty$ for $\zeta \to \infty$, here we can only use the Ai function. In addition, since

$$\frac{2}{3}|\zeta|^{3/2} = \frac{2}{3}k_0 x_c^{-1/2}(x_c - x)^{3/2} = k_0 \int_x^{x_c} (\varepsilon(x))^{1/2} dx, \qquad (7.15)$$

at the plasma boundary, where we can use the asymptotic limit (7.13), the electric field is given by

$$E_{x=0} = 1 + \exp\left[-i2k_0 \int_0^{x_c} (\varepsilon(x))^{1/2} dx + i\frac{\pi}{2}\right], \qquad (7.16)$$

assuming an incident wave with unit amplitude. Thus, apart from a constant, the phase delay of the reflected wave (second term) is given by the geometrical optics approximation (7.6).

For arbitrary permittivity we can still use the above procedure inside a narrow layer of thickness 2δ around the cutoff satisfying the condition

$$\left|\frac{d\varepsilon^2}{dx^2}\right|_c \delta \ll \left|\frac{d\varepsilon}{dx}\right|_c, \qquad (7.17)$$

with the subscript indicating the value at $x = x_c$. Then, if $|\zeta| \geq 5$ at $x = x_c - \delta$, we can join the Airy solution, valid for $|x_c - x| \leq \delta$, with that from the WKBJ approximation [2, 17] and thus recover again the phase delay (7.6), as explained in the next section. This requires $(k_0/|d\varepsilon/dx|_c)^{2/3}|d\varepsilon/dx|_c\delta \geq 5$ (from (7.11)) so that (7.17) becomes

$$\left|\frac{d\varepsilon^2}{dx^2}\right|_c \ll \left|\frac{d\varepsilon}{dx}\right|_c^{4/3}\frac{k_0^{2/3}}{5}, \tag{7.18}$$

which is well satisfied in tokamak plasmas, with the exception of a narrow region near a maximum of the electron density for the O and L cutoffs.

7.3 WKBJ Approximation

The WKBJ approximation — a powerful but simple technique for solving many physical problems — can be applied to the propagation of waves with wavelengths much shorter than the scale of the medium.

Starting from the approximation of geometrical optics, it seems reasonable to take for two independent solutions of (7.10) the functions

$$\exp\left[\pm ik_0 \int \varepsilon^{1/2}(x)dx\right], \tag{7.19}$$

which when inserted into the left-hand side of (7.10) yield

$$\left(\frac{d^2}{dx^2} + k_0^2\varepsilon\right)\exp\left(\pm ik_0 \int \varepsilon^{1/2}(x)dx\right)$$

$$= \pm ik_0\frac{d\varepsilon^{1/2}}{dx}\exp\left(\pm ik_0 \int \varepsilon^{1/2}(x)\,dx\right). \tag{7.20}$$

For (7.19) to be a good approximation, (7.20) must be very small compared to either terms of (7.10), from which we get the condition

$$k_0 \gg \frac{1}{2}\left|\frac{1}{\varepsilon^{3/2}}\frac{d\varepsilon}{dx}\right|. \tag{7.21}$$

If E_1 and E_2 are two independent solutions of (7.10), their Wronskian $W(E_1, E_2) \equiv E_1 dE_2/dx - E_2 dE_1/dx$ must be constant. On the other hand, it is easily shown that the functions (7.19) do not satisfy this condition since their Wronskian is equal to $-2ik_0\varepsilon^{1/2}(x)$. However, if we add to (7.19) an additional factor $\varepsilon^{-1/4}$, the Wronskian becomes independent of x. Note that this factor is similar to the coefficient $\zeta^{-1/4}$ in the asymptotic expansion of the Airy functions. This suggest that

$$\frac{1}{\varepsilon^{1/4}} \exp\left[\pm ik_0 \int \varepsilon^{1/2}(x)\, dx\right] \tag{7.22}$$

provides a better approximation. By inserting these into the left-hand side of (7.10) we get

$$\left[\frac{3}{4\varepsilon}\left(\frac{d\varepsilon^{1/2}}{dx}\right)^2 - \frac{1}{2\varepsilon^{1/2}}\frac{d^2\varepsilon^{1/2}}{dx^2}\right]\frac{1}{\varepsilon^{1/4}}\exp\left[\pm ik_0 \int \varepsilon^{1/2} dx\right]. \tag{7.23}$$

Again, from the comparison of this to the terms of (7.10), we obtain the condition for the validity of the first-order WKBJ approximation

$$\left|\frac{3}{4N^4}\left(\frac{dN}{dx}\right)^2 - \frac{1}{2N^3}\frac{d^2N}{dx^2}\right| \ll k_0^2, \tag{7.24}$$

which must be considered a quantitative definition of the term *slowly varying*. This shows that the approximation breaks down near a cutoff regardless of the value of the refractive index derivatives. It also indicates that the condition is most easily satisfied at high frequencies.

We are now ready to explain how to do the joining of the Airy functions with the WKBJ solution. Around the point $x = x_c - \delta \equiv x_w$ where $\zeta < -5$, the WKBJ function $\varepsilon^{-1/4}\sin[k_0\int_{-\infty}^{x_w}\varepsilon^{1/2}(s)ds]$ should be attached to the Airy function Ai to form the function $A(x)$. This can be done by adding to the WKBJ function a multiplier and a phase so that one of its maxima or minima near x_w coincides with that of Ai. The same multiplier and phase should then be added to $\varepsilon^{-1/4}\cos[k_0\int_{-\infty}^{x_w}\varepsilon^{1/2}(s)ds]$ and attached to Bi at $x = x_w$ to form the function $B(x)$. Finally, for the sake of simpler notations, we may renormalize A and B to unit amplitude in the vacuum region, so that

$$W(A, B) = -k_0. \tag{7.25}$$

7.4 Calculation of Electron Density

The main conclusion from the previous two sections is that the phase ϕ of reflected waves by a plasma cutoff is given under quite general conditions by the approximation of geometric optics (7.6). Thus, its frequency dependence $\phi(\omega)$, which can be measured by sweeping the frequency of the probing wave, can provide a measurement of the electron density. On the other hand, since the phase of the reflected wave can be obtained only relative to the unknown phase of a reference signal, the direct output of reflectometry measurements is $d\phi(\omega)/d\omega$, rather than $\phi(\omega)$. Indeed, this is a quantity with a clear physical meaning since, by performing the derivative with respect to ω of the local dispersion relation (3.9) and recalling (3.14), $d\phi(\omega)/d\omega$ can be cast in the form

$$\frac{d\phi(\omega)}{d\omega} = 2 \int_{x_0}^{x_c(\omega)} \frac{dx}{v_G}, \tag{7.26}$$

clearly indicating that it is the round-trip group delay of the reflected wave. For propagation perpendicular to the magnetic field, the cold plasma approximation gives

$$v_{GO} = c\,N_O,$$

$$v_{GX} = c\,N_X \frac{(1 - X - Y^2)^2}{(1 - X - Y^2)^2 + XY^2}, \tag{7.27}$$

for the group velocity of the ordinary and extraordinary mode, respectively. In ionospheric studies, because of the large distance between the launcher and the reflecting layer, the group delay is obtained directly from the time of flight of short pulses [1,2]. In laboratory plasmas, the group delay is instead obtained from phase measurements. Nevertheless, recent technological advances are beginning to make the ionospheric approach feasible in laboratory plasmas as well.

When $x_c(\omega)$ is a monotonic function of frequency, the electron density profile can be obtained by inverting (7.26). For the O-mode in the cold plasma approximation, this can be done analytically since v_{GO} is only a function of ω and of the local cutoff frequency $\omega = \omega_p$. By using the latter as the independent variable, (7.26) can be written as an Abel integral equation [2], which can be inverted. However, for the X-mode, and in those cases where relativistic corrections must be taken into account, the inversion of (7.26) cannot be performed analytically since the group velocity is an explicit function of position. In these cases, the inversion can be calculated numerically with the following procedure. Let $x_0 = 0$ be the known cutoff position corresponding to the lowest frequency ω_0, and $\omega_1, \omega_2, \ldots, \omega_n$ be the frequencies where the group delay is measured, and define

$$\tau_i = \left(\frac{d\phi(\omega)}{d\omega} \right)_{\omega=\omega_i},$$

$$\phi_i = \sum_{j=1}^{i} \tau_j (\omega_j - \omega_{j-1}),$$

$\qquad(7.28)$

To simplify notations, we assume that the contribution to the group delay of the edge region $x < x_0$ has been subtracted from the measured value of $d\phi(\omega)/d\omega$. Let x_i be the cutoff position of the frequency ω_i, and define $A_{i,j} = \omega_i(N_{i,j} + N_{i,j-1})/2c$, where $1 \le j \le i$, $N_{i,j}$ is the refractive index for $\omega = \omega_i$ and $x = x_j$, and $N_{i,i} = 0$ by definition. Then (7.6) becomes, for all values of i,

$$\phi_i = \sum_{j=1}^{i} A_{i,j}(x_j - x_{j-1}), \qquad(7.29)$$

which in matrix notation can be written as

$$
\begin{bmatrix} \phi_1 \\ \phi_2 \\ \phi_3 \\ \vdots \\ \phi_n \end{bmatrix}
=
\begin{bmatrix}
A_{1,1} & 0 & 0 & \cdots \\
A_{2,1} - A_{2,2} & A_{2,2} & 0 & \cdots \\
A_{3,1} - A_{3,2} & A_{3,2} - A_{3,3} & A_{3,3} & \cdots \\
\vdots & \vdots & \vdots & \ddots \\
A_{n,1} - A_{n,2} & A_{n,2} - A_{n,3} & A_{n,3} - A_{n,4} & \cdots
\end{bmatrix}
\begin{bmatrix} x_1 \\ x_2 \\ x_3 \\ \vdots \\ x_n \end{bmatrix},
$$

$\qquad(7.30)$

or more concisely

$$\boldsymbol{\phi} = \mathbf{M} \cdot \mathbf{x}. \tag{7.31}$$

For the O-mode in the cold plasma approximation, the matrix $\mathbf{M}$ is not an explicit function of x_i, and thus the vector $\mathbf{x}$ can be easily obtained by inverting (7.31)

$$\mathbf{x} = \mathbf{M}^{-1} \cdot \boldsymbol{\phi}. \tag{7.32}$$

For the general case, from (7.31) we get

$$
\begin{aligned}
x_1 &= \phi_1/A_{1,1} \\
x_2 &= [x_1(A_{2,2} - A_{2,1}) + \phi_2]/A_{2,2} \\
x_3 &= [x_1(A_{3,2} - A_{3,1}) + x_2(A_{3,3} - A_{3,2}) + \phi_3]/A_{3,3} \\
&\vdots \\
x_n &= [x_1(A_{n,2} - A_{n,1}) + x_2(A_{n,3} - A_{n,2}) \\
&\quad + \cdots + x_{n-1}(A_{n,n} - A_{n,n-1}) + \phi_n)]/A_{n,n},
\end{aligned}
\tag{7.33}
$$

that, since x_i is given in terms of x_j with $j < i$, represents the solution of our problem for any type of wave polarization. Once the values of x_i have been determined, the electron density is easily obtained from the known dependence of the cutoff frequencies on plasma parameters. The precision then depends on the accuracy with which the group delays τ_i are measured. In most reflectometry schemes, the received signal is downshifted at a lower frequency ω_s and the group delay is obtained from phase measurements. These may be considered equivalent to the measurement of the zero-crossing time t of the received waveform. In the presence of noise with a large signal-to-noise power ratio (S/N), the root mean square error is then given by $\delta t = 1/\omega_s(2S/N)^{1/2}$ [18], so that the uncertainty in the position of the cutoff is

$$\delta x \approx \frac{c}{2\omega_s(2S/N)^{1/2}}. \tag{7.34}$$

This equation shows the importance of operating at large frequencies, which allows a better statistical averaging of the results as well.

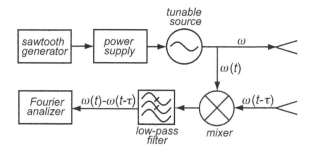

Figure 7.6. Block diagram of the FMCW reflectometer.

Several methods are available for reflectometry measurements of plasma density profiles, all based on the measurement of the round trip group delay of the probing beam. One of the first to be proposed [19] is a modification of the FMCW radar technique [18], where the round trip delay (τ) is converted to an easily measured beat frequency.

A schematic of the FMCW reflectometer is shown in Fig. 7.6, where a wave oscillator generates a time-dependent frequency $\omega(t)$, which is sent to both the plasma and the reference port of the receiving mixer. Since the downshifted beat frequency (ω_B) is

$$\omega_B = \omega(t) - \omega(t - \tau) \approx \tau \frac{d\omega}{dt},$$

by choosing $d\omega/dt$ nearly constant in magnitude, we obtain the desired simple dependence of the round trip delay on the beat frequency.

As we shall see in the remaining of this chapter, the short-scale density fluctuations of magnetized plasmas can be a source of strong noise. In order to overcome this problem, modern FMCW reflectometers use solid-state sources that can be swept in a time shorter than the characteristic time of turbulence ($\sim 10\,\mu s$), allowing the measurements to be made on a *quasi-frozen* plasma [20, 21].

Other types of reflectometers for density measurements are the dual frequency differential phase systems, where the group delay is obtained by launching simultaneously two waves with a frequency separation small enough ($\sim 100\,$MHz) to make their cutoffs distance shorter than the turbulence correlation length. The two waves are created by either mixing a microwave with the signal from two crystal oscillators [22] or by amplitude modulation of a single wave [23]. The density profile is obtained by keeping constant the frequency separation while the signals are swept across a frequency band. Compared to the FMCW scheme, this type of reflectometer is seriously affected by spurious reflections.

Finally, the wave group delay can be measured using pulsed radar techniques [24], the difficulty being the very short pulse lengths that are required for obtaining a good spatial resolution in fusion plasmas.

7.5   Fluctuations Measurements

As mentioned in the introduction, microwave reflectometry is also a tool for the detection of plasma fluctuations, even though, as we shall see in the next sections, the extraction of quantitative information from measured signals is often very difficult.

In the presence of density fluctuations, the interpretation of reflectometry data is relatively simple when the plasma permittivity varies only along the direction of propagation of the probing wave. This can be easily seen in a plane-stratified geometry with the probing wave propagating with the ordinary mode along the x-axis, and by taking a wave permittivity in the form

$$\varepsilon = \varepsilon_0(x) + \tilde{\varepsilon}(x), \tag{7.35}$$

with $|\tilde{\varepsilon}(x)| \ll 1$. We assume that $d\tilde{\varepsilon}(x)/dx \neq 0$ at the cutoff $(x = x_c)$, and that the thickness of the evanescent region behind it is many free space wavelengths thick, so that tunneling effects may be ignored. Then, we proceed by making the *Ansatz*, to be verified *a posteriori*, that [25, 26]

$$E = E_0 + \sum_{n>0} E_n, \tag{7.36}$$

with $|E_0| \sim O(1)$ and $|E_{n>0}| \sim O(\tilde{\varepsilon}^n)$, so that the two lowest terms of (7.10) are

$$\frac{d^2 E_0}{dx^2} + k_0^2 \varepsilon_0(x) E_0 = 0 \tag{7.37}$$

and

$$\frac{d^2 E_1}{dx^2} + k_0^2 \varepsilon_0(x) E_1 = -k_0^2 \tilde{\varepsilon}(x) E_0(x). \tag{7.38}$$

As explained in Sec. 7.3, two independent solutions of the first of these equations can be obtained as a combination of the Airy functions and the WKBJ approximation. The second equation may be solved using the method of *Variation of Parameters* [27], a powerful technique that can be easily explained for the case under discussion.

Given the non-homogeneous equation

$$\frac{d^2 u}{dx^2} + f(x) u = g(x), \tag{7.39}$$

let $u_1(x)$ and $u_2(x)$ be two independent solutions of the corresponding homogenous equation

$$\frac{d^2 u}{dx^2} + f(x) u = 0, \tag{7.40}$$

with Wronskian

$$W(u_1, u_2) = u_1 \frac{du_2}{dx} - u_2 \frac{du_1}{dx}. \tag{7.41}$$

We now look for a solution of (7.39) in the form

$$u = a u_1 + b u_2, \tag{7.42}$$

where a and b are two functions of x to which we are free to impose one constraint. For this we choose

$$u_1 \frac{da}{dx} + u_2 \frac{db}{dx} = 0, \tag{7.43}$$

so that

$$\frac{du}{dx} = a \frac{du_1}{dx} + b \frac{du_2}{dx}. \tag{7.44}$$

It follows that

$$\frac{d^2 u}{dx^2} = \frac{du_1}{dx} \frac{da}{dx} + \frac{du_2}{dx} \frac{db}{dx} + a \frac{d^2 u_1}{dx^2} + b \frac{d^2 u_2}{dx^2}. \tag{7.45}$$

Substituting this in (7.39) and using (7.40) gives

$$\frac{du_1}{dx}\frac{da}{dx} + \frac{du_2}{dx}\frac{db}{dx} = g(x), \tag{7.46}$$

which together with (7.43) yields

$$\frac{da}{dx} = -\frac{u_2 g}{W(u_1, u_2)}, \qquad \frac{db}{dx} = \frac{u_1 g}{W(u_1, u_2)}, \tag{7.47}$$

an thus (7.42) becomes

$$u(x) = -u_1(x)\int_{x_1}^{x}\frac{u_2(s)g(s)}{W(u_1, u_2)}ds - u_2(x)\int_{x}^{x_2}\frac{u_1(s)g(s)}{W(u_1, u_2)}ds, \tag{7.48}$$

where x_1 and x_2 are arbitrary constants to be determined from the boundary conditions. For the case of interest here, $W(u_1, u_2)$ is constant for (7.40) and thus may be taken outside of the integrals of (7.48).

The solution of (7.38) may then be cast in the form

$$E_1(x) = \frac{k_0^2}{W(y_1, y_2)}\left\{ y_1(x)\int_{-\infty}^{x} y_2(s)\tilde{\varepsilon}(s)E_0(s)ds \right.$$

$$\left. + y_2(x)\int_{x}^{\infty} y_1(s)\tilde{\varepsilon}(s)E_0(s)ds \right\}, \tag{7.49}$$

where $y_1(x)$ and $y_2(x)$ are two independent solutions of (7.37) with Wronskian $W(y_1, y_2)$. For these we may take a linear combination of the two functions A and B of Sec. 7.3, which, as explained there, are obtained by joining the Airy functions Ai and Bi to the WKBJ solutions. The coefficients of the linear combination are obtained from the boundary conditions of $y_1(x)$ and $y_2(x)$. Thus, since $E_1(x) \to 0$ for $x \to \infty$ and $E_1(x) \to \exp[-ik_0 x]$ for $x \to -\infty$, we get $y_1(x) = A(x)$ and $y_2(x) = B(x) - iA(x)$ and thus, since $W(y_1, y_2) = W(A, B)$, (7.49) becomes

$$E_1(x) = -k_0\left\{ A(x)\int_{-\infty}^{x} \tilde{\varepsilon}(s)[B(s) - iA(s)]E_0(s)ds \right.$$

$$\left. + [B(x) - iA(x)]\int_{x}^{\infty} \tilde{\varepsilon}(s)A(s)E_0(s)ds \right\}. \tag{7.50}$$

Finally, in the vacuum region we obtain (apart from a constant phase)

$$E_1(x) = k_0 \exp[-ik_0 x] \int_0^\infty \tilde{\varepsilon}(s) A(s) E_0(s) ds, \qquad (7.51)$$

which, since $E_0(x) = A(x)$, can be cast in the form

$$E_1(x) = k_0 \exp[-ik_0 x] \int_0^\infty \tilde{\varepsilon}(s) A^2(s) ds. \qquad (7.52)$$

From this, by taking the backward wave in the form $\exp(ik_0 x + i\tilde{\phi})$, we get the contribution of fluctuations to the phase of the reflected wave

$$\tilde{\phi} = k_0 \int_0^\infty \tilde{\varepsilon}(s) A^2(s) ds. \qquad (7.53)$$

Away from the cutoff, where A is given by the WKBJ approximation, the coefficient of $\tilde{\varepsilon}$ in (7.53) performs an average over a distance $\delta x = \pi/k_0 \varepsilon_0^{1/2}$, and thus the scattered field originates from fluctuations where the spatial variation of A^2 matches that of the density perturbation (Bragg condition). Near the cutoff the scattered signal is instead strongly weighted by fluctuations with a wavelength longer that the size of the last lobe of $A^2(\zeta)$ with width $\delta\zeta \approx 3$ (Fig. 7.5), i.e., $\delta x \approx 3(k_0 L_\varepsilon)^{1/3}/k_0$ and $L_\varepsilon = 1/|d\varepsilon_0/dx|_{x=x_c}$. For these fluctuations we can use the approximation of geometrical optics, where by expanding $\varepsilon^{1/2}$ to first order in $\tilde{\varepsilon}$ we obtain

$$\tilde{\phi} = k_0 \int_0^{x_c} \frac{\tilde{\varepsilon}}{\varepsilon^{1/2}} ds. \qquad (7.54)$$

This can be used for obtaining the power spectrum $\Gamma_\varepsilon(k_x)$ of $\tilde{\varepsilon}$ from the power spectrum $\Gamma_\phi(k_x)$ of $\tilde{\phi}$. For a turbulence that is homogeneous perpendicularly to the magnetic field, we have

$$\langle \tilde{\varepsilon}(x_2)\tilde{\varepsilon}(x_1) \rangle = \int \Gamma_\varepsilon(k_x) \exp[ik_x(x_2 - x_1)] dk_x,$$

$$\langle \tilde{\phi}(x_{c2})\tilde{\phi}(x_{c1}) \rangle = \int \Gamma_\phi(k_x) \exp[ik_x(x_{c2} - x_{c1})] dk_x, \qquad (7.55)$$

where x_{c1} and x_{c2} are two cutoff layers and the bracket $\langle \rangle$ indicates the ensemble average. By taking $\varepsilon(x) \approx (x_c - x)/L_\varepsilon$, since most of the contribution to $\tilde{\phi}$ comes from a narrow region around the cutoff, from (7.54) and (7.55) we obtain

$$\langle \tilde{\phi}(x_{c2})\tilde{\phi}(x_{c1}) \rangle$$

$$\approx k_0^2 L_\varepsilon \left\langle \int_0^{x_{c2}} \frac{\tilde{\varepsilon}(x_2)}{(x_{c2}-x_2)^{1/2}}dx_2 \int_0^{x_{c1}} \frac{\tilde{\varepsilon}(x_1)}{(x_{c1}-x_1)^{1/2}}dx_1 \right\rangle$$

$$= k_0^2 L_\varepsilon \int \Gamma_\varepsilon(k_x) \exp[ik_x(x_{c2}-x_{c1})]$$

$$\times \left\{ \int_0^{x_{c2}} \frac{\exp[-ik_x(x_{c2}-x_2)]}{(x_{c2}-x_2)^{1/2}}dx_2 \right.$$

$$\times \left. \int_0^{x_{c1}} \frac{\exp[ik_x(x_{c1}-x_1)]}{(x_{c1}-x_1)^{1/2}}dx_1 \right\} dk_x$$

and thus

$$\frac{\Gamma_\phi(k_x)}{\Gamma_\varepsilon(k_x)} \approx k_0^2 L_\varepsilon \left| \int_0^{x_c} \frac{\exp[ik_x(x_c-x)]}{(x_c-x)^{1/2}}dx \right|^2$$

$$= \frac{k_0^2 L_\varepsilon}{k_x} \left| \int_0^{k_x x_c} \frac{\exp[i\alpha]}{\alpha^{1/2}}d\alpha \right|^2 = 4\frac{k_0^2 L_\varepsilon}{k_x} \left| \int_0^{\sqrt{k_x x_c}} \exp[i\beta^2]d\beta \right|^2$$

$$= 2\pi \frac{k_0^2 L_\varepsilon}{k_x} \left| \int_0^{\sqrt{2k_x x_c/\pi}} \exp[i\pi\gamma^2/2)]d\gamma \right|^2,$$

where the integral can be expressed in terms of the Fresnel integrals [16]

$$C(z) = \int_0^z \cos(\pi\gamma^2/2)d\gamma, \quad S(z) = \int_0^z \sin(\pi\gamma^2/2)d\gamma,$$

with $z = \sqrt{2k_x x_c/\pi}$. Since for the plasmas of interest in fusion research $z \gg 1$ and the asymptotic value of Fresnel integrals is 0.5, we finally get

$$\Gamma_\phi(k_x) = \pi \frac{k_0^2 L_\varepsilon}{k_x} \Gamma_\varepsilon(k_x). \tag{7.56}$$

This can be expressed in terms of the power spectrum Γ_n of $\tilde{n}_e/n_e$ and of the density scale length $L_n = n_e/(dn_e/dx)$ as

$$\Gamma_\phi(k_x) = \pi M \frac{k_0^2 L_n}{k_x} \Gamma_n(k_x), \tag{7.57}$$

where M is the value of $n_e(\partial\varepsilon/\partial n_e)$ near the cutoff. For perpendicular propagation to the magnetic field, it is easy to show that $M \approx 1$ for the ordinary mode and $M \approx 2$ for the extraordinary mode.

In conclusion, for long wavelengths one-dimensional turbulent fluctuations, the power spectrum of density fluctuations can be obtained from the power spectrum of the phase of reflected waves. This can be measured by performing correlation measurements using several probing waves with closely spaced cutoff layers.

7.6 Multidimensional Turbulent Fluctuations

The interpretation of reflectometry becomes considerably more complicated for broadband multidimensional fluctuations, as in tokamak plasmas where the wave refractive index varies not only along the direction of propagation of the probing wave, as assumed before, but perpendicularly as well. The result is that even for fluctuations with a wavelength sufficiently long to make geometrical optics applicable, the measured wave field cannot be described as a specular reflection of the probing wave, as in the case of one-dimensional fluctuations, and more important, its properties can differ substantially from those of plasma fluctuations. This can be seen by taking the dielectric tensor $\boldsymbol{\varepsilon} = \boldsymbol{\varepsilon}_0(x) + \tilde{\boldsymbol{\varepsilon}}(x, y)$ in the system of Cartesian coordinates (x, y, z) where the probing wave is launched in the x-direction and the magnetic field is in the z-direction. Since $\boldsymbol{\varepsilon}$ does not depend on z, we can assume the wave electric field to be independent of z as well. From the wave equation (7.7) we get

$$\frac{\partial^2 E_x}{\partial x^2} + \frac{\partial^2 E_x}{\partial y^2} + \frac{\omega^2}{c^2}(\varepsilon_{xx}E_x + \varepsilon_{xy}E_y) - \frac{\partial}{\partial x}\nabla \cdot \mathbf{E} = 0,$$

$$\frac{\partial^2 E_y}{\partial x^2} + \frac{\partial^2 E_y}{\partial y^2} + \frac{\omega^2}{c^2}(\varepsilon_{yx}E_x + \varepsilon_{yy}E_y) - \frac{\partial}{\partial y}\nabla \cdot \mathbf{E} = 0, \qquad (7.58)$$

$$\frac{\partial^2 E_z}{\partial x^2} + \frac{\partial^2 E_z}{\partial y^2} + \frac{\omega^2}{c^2}\varepsilon_{zz}E_z = 0$$

showing that E_z is fully independent of the other two wave components. Then, since the wave field can always decomposed into two waves with mutually perpendicular directions of polarization, we can take one of the latter to be the z-axis and consider for simplicity only the solution of the last of (7.58). For this, we consider a perturbed permittivity of the type (dropping the subscript from ε_{zz})

$$\tilde{\varepsilon}(x,y) = \tilde{\varepsilon}(x)\exp(ipy) \qquad (7.59)$$

and expand the solution to first order in $\tilde{\varepsilon}$ as in Sec. 7.5. The zeroth-order equation is identical to (7.37). For the first order we search for solutions of the form

$$E_1 = E_1(x)\exp(ipy) \qquad (7.60)$$

which satisfies the equation

$$\frac{d^2 E_1}{dx^2} + [k_0^2 \varepsilon_0(x) - p^2]E_1 = -k_0^2 \tilde{\varepsilon}(x)E_0(x) \qquad (7.61)$$

that is similar to (7.38). Thus, the solution of our problem can proceed as in the previous section.

In summary, a small sinuously perturbation of the cutoff in the y-direction has the effect of producing a side lobe of the reflected wave with the same wave number of the perturbation in the y-direction, which therefore propagates at a different direction from that of specular reflection. In other words, the cutoff behaves as a grating. This offers the possibility of obtaining the propagation velocity of the plasma perturbation from the Doppler shift of the side lobe — the *Doppler Reflectometry* technique [28, 29].

The difficulty arises in the case of a broadband turbulence, where each spectral component will produce its own diffracted waves that, if allowed to propagate freely to the observation point, will create a chaotic interference pattern with statistical features that are unrelated to those of the turbulence under investigation. This can be explained with the following arguments.

In plasmas where the density is perturbed by a short-scale broadband turbulence, the phase of the reflected wave is the cumulative result of many random contributions, and thus it is reasonable to assume that the phase modulation $\tilde{\phi}$ near the cutoff is a normal random variable with mean $\langle \tilde{\phi} \rangle = 0$, variance $\sigma_\phi^2 \equiv \langle \tilde{\phi}^2 \rangle$ and autocorrelation

$$\gamma_\phi(\xi) \equiv \langle \tilde{\phi}_1(y) \tilde{\phi}_2(y + \xi) \rangle / \sigma_\phi^2 \qquad (7.62)$$

(with brackets $\langle \rangle$ representing the ensemble average).

We now introduce the characteristic function of $\tilde{\phi}$ and of the phase difference $\Delta \tilde{\phi} = \tilde{\phi}_1 - \tilde{\phi}_2$, given by [30, 31]

$$\begin{aligned} M_\phi(\omega) &\equiv \langle \exp[i\omega\tilde{\phi}] \rangle \\ &= \frac{1}{\sigma_\phi \sqrt{2\pi}} \int_{-\infty}^{\infty} \exp[i\omega\tilde{\phi}] \exp[-\tilde{\phi}^2/2\sigma_\phi^2] d\tilde{\phi} \\ &- \exp[-\sigma_\phi^2 \omega^2/2] \end{aligned} \qquad (7.63)$$

and

$$\begin{aligned} M_{\Delta\phi}(\omega) &\equiv \langle \exp[i\omega\Delta\tilde{\phi}] \rangle \\ &= \frac{1}{\sigma_\phi \sqrt{2\pi}} \int_{-\infty}^{\infty} \exp[i\omega\Delta\tilde{\phi}] \exp[-\Delta\tilde{\phi}^2/2\sigma_{\Delta\phi}^2] d\tilde{\phi} \\ &= \exp[-\sigma_{\Delta\phi}^2 \omega^2/2], \end{aligned} \qquad (7.64)$$

respectively. Since

$$\sigma_{\Delta\phi}^2 = \langle \Delta\tilde{\phi}^2 \rangle = \langle \tilde{\phi}_1^2 \rangle + \langle \tilde{\phi}_2^2 \rangle - 2\langle \tilde{\phi}_1 \tilde{\phi}_2 \rangle = 2(\langle \tilde{\phi}^2 \rangle - \sigma_\phi^2 \gamma_\phi), \qquad (7.65)$$

(7.64) becomes

$$M_{\Delta\phi}(\omega) = \exp[-\sigma_\phi^2(1 - \gamma_\phi)\omega^2]. \qquad (7.66)$$

From this we obtain the first moment of the wave electric field (which can be interpreted as the amplitude of the coherent specular reflection)

$$\langle E \rangle = M_\phi(1) = \exp[-\sigma_\phi^2/2], \qquad (7.67)$$

and thus it is a decreasing function of σ_ϕ. For the second moment we get

$$\langle E_1 E_2^* \rangle = M_{\Delta\phi}(1) = \exp[-\sigma_\phi^2(1 - \gamma_\phi)], \qquad (7.68)$$

which proves that the signal correlation length is also a decreasing function of σ_ϕ. In particular, for $\sigma_\phi \gg 1$, taking $\gamma_\phi(\xi) = \exp[-(\xi/\Delta)^2]$ and expanding to the second order in ξ, we obtain

$$\langle E_1 E_2^* \rangle = \exp[-(\sigma_\phi\xi/\Delta)^2], \qquad (7.69)$$

so that in the presence of large fluctuations, the spectrum of reflected waves becomes broader than the spectrum of $\tilde{\phi}$ by a factor σ_ϕ and consequently broader than the spectrum of plasma fluctuations.

Based on these arguments, a *phase screen model* of reflectometry in the presence of two-dimensional turbulent fluctuations has been proposed [32], where the field of scattered waves arises near the cutoff from the phase modulation of the probing wave with the magnitude given by geometrical optics. To an observer at a distance from the cutoff, the reflected waves appear to originate at a virtual location behind the real cutoff because of refraction in a medium with refractive index smaller than one. After reflection, the electromagnetic field separates into a wave propagating along the direction of specular reflection, and into a group of scattered waves propagating in different directions. The amplitude of the former decreases quickly to a small level as the variance σ_ϕ^2 of the phase modulation becomes larger than one, as implied by (7.67). This phenomenon is clearly illustrated in Fig. 7.7, which shows the measured spectrum of reflected waves in two plasmas of the Tokamak Fusion Test Rector (TFTR), one with a relatively low central ion temperature of 5 keV and a low level of fluctuations, the other with a central ion temperature of 20 keV and large turbulent fluctuations. The spectra $S_E(\omega)$ are those of the common power between two closely spaced reflectometers with reflecting layers separated by a distance of the order of the probing wavelength. This explains why their spectral coherence (γ_E) is close to unity over

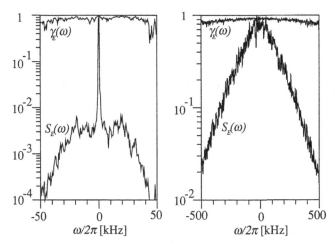

Figure 7.7. Frequency power spectrum $S_E(\omega)$ of reflected waves in plasmas with central ion temperatures of 5 (left) and 20 (right) keV. Spectra are from the common power of two closely spaced reflectometer channels with a cutoff separation of $\approx 2\pi/k_0$ and spectral coherence $\gamma_E(\omega)$. (With permission from [32]).

the entire spectral range. Nevertheless, the spike at $\omega = 0$ in the frequency spectrum of the case with low fluctuations, which represents the amplitude of the coherent specular reflection, has totally disappeared in the high fluctuation case, as predicted by (7.67).

As suggested by the proposed model of reflectometry, the transition from weak to strong scattering should be accompanied by a spectral broadening, which is clearly shown in Fig. 7.7. This broadening is due to the fact that the spectral width of scattered waves increases with the amplitude of fluctuations and becomes a factor of σ_ϕ larger than the spectrum of $\tilde{\phi}$ at the cutoff (7.69). Thus, if $|\delta k_y|$ is the spectral width of plasma fluctuations in the y-direction, the scattered waves will spread over a range of wave numbers $|\delta k_y| = \sigma_\phi^2 |\delta k_y|$, so that an observer at the plasma edge will sample a chaotic interference pattern with large amplitude variations and random phases. This suggests that the amplitude of the received signal S should follow the distribution derived by Rice [33, 34] for a signal containing a sinusoidal component — the un-scattered wave — and a Gaussian noise with both its real and imaginary parts as independent random variables. Thus, if $S = \alpha + i\beta$, the signal distribution is

$$P(\alpha, \beta) = \frac{1}{2\pi\sigma^2} \exp[-(\alpha - \alpha_0)^2/2\sigma^2 - \beta^2/2\sigma^2], \qquad (7.70)$$

where σ^2 is the variance of the Gaussian noise, and α_0 is the amplitude of the sinusoidal component of S, which in our case is equal to $\exp[-\sigma_\phi^2/2]$ because of (7.67). From this, we easily get

$$\langle \alpha^2 \rangle = \frac{1}{\sqrt{2\pi}\sigma} \int_{-\infty}^{+\infty} \alpha^2 \exp[-(\alpha - \alpha_0)^2/2\sigma^2]d\alpha = \sigma^2 + \alpha_0^2, \quad (7.71)$$

and thus

$$\langle \alpha^2 + \beta^2 \rangle = 2\sigma^2 + \alpha_0^2 = 1, \quad (7.72)$$

where we assume a probing wave with unit amplitude. In the polar system of coordinates (ρ, θ) with $\alpha = \rho \cos\theta$ and $\beta = \rho \sin\theta$, (7.70) can be written as

$$P(\alpha, \beta)d\alpha\, d\beta = \frac{1}{2\pi\sigma^2} \exp[-(\rho^2 + \alpha_0)^2/2\sigma^2 + \rho\alpha_0 \cos\theta/\sigma^2]\rho d\rho d\theta,$$
$$(7.73)$$

whose integration in θ gives the Rice distribution of the signal amplitude

$$P(\rho) = \frac{\rho}{\sigma^2} \exp[-(\rho^2 + \alpha_0^2)/2\sigma^2]I_0(\rho\alpha_0/\sigma^2), \quad (7.74)$$

where I_0 is the modified Bessel function of order zero [16].

Figure 7.8 illustrates two examples taken from TFTR plasmas [35], where the measured amplitude distribution of reflected waves in the far-field region is compared with the Rice distribution that best fit the experimental data, giving a σ_ϕ of 0.4 rad for wave reflection from the plasma center (a), where fluctuations are small, and 0.75 rad for reflection from an outer region with a larger level of turbulence. The excellent agreement of the experimental signal distribution with the Rice distribution (7.74) is a strong indication of the Gaussian distribution of scattered waves. Hence the conclusion that the spectrum of plasma fluctuations cannot be inferred from the phase of measured signals. Nevertheless, σ_ϕ is an important piece of information on the strength of plasma turbulence. Unfortunately, even σ_ϕ cannot be obtained from reflectometry measurements when

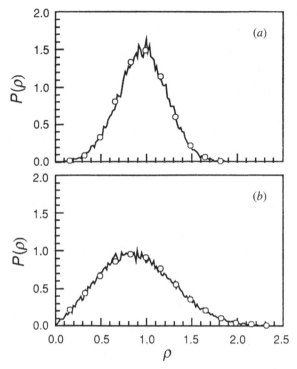

Figure 7.8. Amplitude distribution of reflected waves in a TFTR plasma. Solid line is from an X-mode reflectometer with phase quadrature detection. Circles are from the best fit of experimental data with a Rice distribution having $\sigma_\phi = 0.4$ rad (a) and $\sigma_\phi = 0.75$ rad (b). (With permission from [35]).

its value exceeds $\sim$3 rad, for which $\alpha_0 \ll 1$ and $\sigma^2 \approx 1/2$ and the Rice distribution becomes the Rayleigh distribution $P(\rho) = 2\rho \exp[-\rho^2]$ (Fig. 7.9), which does not depend on σ_ϕ.

In conclusion, the spectrum of large broadband plasma fluctuations cannot be obtained from the spectrum of reflected waves when allowed to propagate freely to the detection plane and form a chaotic interference pattern. The possibility of overcoming this difficulty was discussed in a series of papers [36–38], where it was shown that the interference of reflected waves could be prevented by using a wide aperture optical system for creating an image of the cutoff onto an array of phase sensitive detectors. This should allow a measurement of the correlation function γ_ϕ and provide information on the

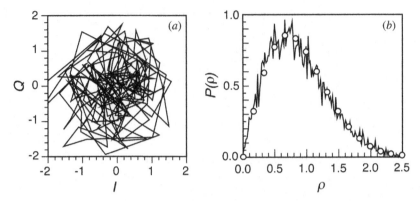

Figure 7.9.   Amplitude of reflected waves in a TFTR plasma with large turbulent fluctuations. (a) in-phase (I) and quadrature (Q) components of reflected signal in a 200 ms time window showing a random distribution; (b) probability distribution of signal amplitude $\rho = (I^2 + Q^2)^{1/2}$ showing a Rayleigh distribution (circles). (With permission from [36]).

structure of plasma density fluctuations. This scheme — *microwave-imaging reflectometry* — was the result of an extensive series of numerical simulations that are summarized in the next section.

7.7 Numerical Simulation

For a numerical simulation of the effect of turbulent fluctuations in reflectometry, we assume that the density of a cylindrically symmetric plasma is perturbed by a field of two-dimensional density fluctuations, which in the system of cylindrical coordinates (r, θ, z) have the spatial distribution

$$\frac{\tilde{n}(r,\theta)}{n(r)} = \sum_{p=1}^{P} \sum_{q=1}^{Q} \delta_{pq} \cos(p\kappa r + \varphi_{pq}) \cos(q\theta), \qquad (7.75)$$

consisting of $P \times Q$ discrete components with radial wave number $k_r = p\kappa$ (where κ is a constant), poloidal number q, random phase φ_{pq} and amplitudes δ_{pq}. For the latter we take the distribution

$$\delta_{pq}^2 \propto p \exp[-(p\kappa/\kappa_r)^2 - (q/q_0)^2], \qquad (7.76)$$

where $\kappa_r = \kappa P/2$ and q_0 represent the spectral width in the radial and θ directions, respectively. In the following, we will indicate with $\kappa_\theta = q_0/r_c$ the width of the spectrum of poloidal wave numbers at the cutoff $(r = r_c)$.

The probing wave is launched from free space at $r = r_0$ (the plasma radius) with a frequency of 75 GHz and the ordinary mode of propagation (electric field in the z-direction) and with the Gaussian amplitude profile

$$E_0(\theta) = \exp[-(\theta/\theta_0)^2], \tag{7.77}$$

where $\theta_0 \ll \pi$ is a constant.

A periodic function $F(\theta)$ with period 2π that is equal to (7.77) when $-\pi < \theta < \pi$ is given by

$$F(\theta) = \sum_{m=-\infty}^{m=\infty} g_m e^{-im\theta}, \tag{7.78}$$

with

$$g_m = \frac{1}{2\pi} \int_{-\pi}^{+\pi} e^{-(\theta/\theta_0)^2} e^{im\theta} d\theta.$$

For $\theta_0 \ll \pi$, we obtain

$$g_m \approx \frac{1}{2\pi} \int_{-\infty}^{+\infty} e^{-(\theta/\theta_0)^2} e^{im\theta} d\theta = \frac{\theta_0}{2\sqrt{\pi}} e^{-(m\theta_0/2)^2},$$

so that (7.78) can be expressed as

$$E_0(\theta) \approx \frac{\theta_0}{2\sqrt{\pi}} \sum_{m=-\infty}^{m=+\infty} e^{-(m\theta_0/2)^2} e^{im\theta}. \tag{7.79}$$

The total wave amplitude is expressed as the sum

$$E(r,\theta) = \sum_{n=-N}^{N} c_n E_n(r,\theta), \tag{7.80}$$

with $N \gg Q$ (to be determined), where the functions $E_n(r, \theta)$ are $2N + 1$ independent solutions of the wave equation

$$\nabla^2 E + k_0^2 \varepsilon E = 0 \qquad (7.81)$$

with permittivity

$$\varepsilon = \varepsilon_0 + (\varepsilon_0 - 1)\frac{\tilde{n}}{n}, \qquad (7.82)$$

where $\varepsilon_0 = 1 - (\omega_p/\omega)^2$ and $\omega_p^2 = 4\pi n e^2/m_e$. Inserting into the wave equation the functions E_n, which are cast in the form

$$E_n(r, \theta) = \sum_{m=-N}^{N} f_{mn}(r)e^{im\theta}, \qquad (7.83)$$

and performing a Fourier expansion in θ yield the system of $2N + 1$ ordinary differential equations

$$\frac{d^2 f_{mn}}{dr^2} + \frac{1}{r}\frac{df_{mn}}{dr} + k_0^2(\varepsilon_0 - \alpha_m^2)f_{mn} + k_0^2(\varepsilon_0 - 1)$$

$$\times \sum_{p=1}^{P}\sum_{q=1}^{Q}\left[\frac{\delta_{pq}}{2}\cos(p\kappa r + \varphi_{pq})(f_{(m-q)n} + f_{(m+q)n})\right] = 0$$

$$(m = -N, -N+1, \ldots, N-1, N), \qquad (7.84)$$

with $\alpha_m = m/k_0 r$. The closure is obtained by setting to zero all terms $f_{(m\pm q)n}$ with $|m \pm q| > N$.

The coefficients c_n in (7.80) are obtained by imposing the condition that the wave field at the launching radius $r = r_0$ is the sum of the incoming wave (7.79) and an outgoing reflected wave that we write as

$$E_r(\theta) = \sum_{m=-N}^{N} a_m e^{im\theta}. \qquad (7.85)$$

From this we get a first set of $2N + 1$ equations

$$\sum_{n=-N}^{N} f_{mn}(r_0)c_n - a_m = g_m \quad (m = -N, -N+1, \ldots, N-1, N).$$

$$(7.86)$$

A second set of equations can be derived from

$$E_f(r,\theta) = \sum_{m=-N}^{N} g_m \frac{H_m^{(2)}(k_0 r)}{H_m^{(2)}(k_0 r_0)} e^{im\theta} \qquad (7.87)$$

and

$$E_b(r,\theta) = \sum_{m=-N}^{N} a_m \frac{H_m^{(1)}(k_0 r)}{H_m^{(1)}(k_0 r_0)} e^{im\theta}, \qquad (7.88)$$

where $H_m^{(1)}(z) = J_m(z) + iY_m(z)$ and $H_m^{(2)}(z) = J_m(z) - iY_m(z)$ are the *Hankel* functions [16] of order m, which satisfy the recurrence relation

$$2F_m'(z) = F_{m-1}(z) - F_{m+1}(z).$$

Since these are the forward and backward solutions of the wave equation in free space, they must coincide with (7.79) and (7.85) at $r = r_0$. From the r-derivatives we obtain a second set of equations

$$\sum_{n=-N}^{N} f_{mn}'(r_0)c_n - a_m \frac{k_0(dH_m^{(1)}(z)/dz)_{z=k_0 r_0}}{H_m^{(1)}(k_0 r_0)}$$

$$= g_m \frac{k_0(dH_m^{(2)}(z)/dz)_{z=k_0 r_0}}{H_m^{(2)}(k_0 r_0)}$$

$$(m = -N, -N+1, \ldots, N-1, N), \qquad (7.89)$$

which together with (7.86) determines the values of a_m and c_n.

Finally, the integer N must be chosen to make the results significantly unchanged by any increase in its value. In the present simulation, N is in the range of 200–250 together with $P = 21$ and $Q = 101$.

The following numerical results were obtained using the density profile of Fig. 7.10 where the ordinary cutoff is at $r = 40\,\text{cm}$ for the probing frequency of 75 GHz.

Figure 7.11 displays the contour plot of the backward field amplitude E_b for $\sigma_n \equiv \langle \tilde{n}^2/n^2 \rangle = 1.0 \times 10^{-2}$, $\kappa_r = 1.0\,\text{cm}^{-1}$, $\kappa_\theta = 0.5\,\text{cm}^{-1}$ $(q_0 = 20)$ and $\theta_0 = 40°$. This plot illustrates how plasma fluctuations break the pattern of E_b into several striations that seems to originate

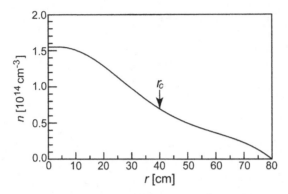

Figure 7.10.   Plasma density profile; the cutoff is at $r = 40$ cm for a wave with the frequency of 75 GHz and the ordinary mode.

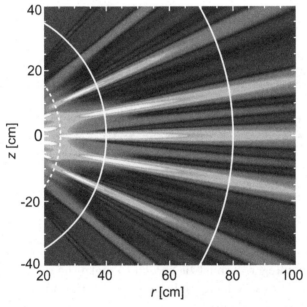

Figure 7.11.   (Color online) Contour plots of $|E_b|$ for $\sigma_n = 1.0 \times 10^{-2}$, $\kappa_r = 1.0\,\text{cm}^{-1}$, $\kappa_\theta = 0.5\,\text{cm}^{-1}$ and $\theta_0 = 40°$ (concentric circles from big to small are plasma boundary, cutoff and virtual cutoff). (With permission from [38]).

from a location behind the cutoff — a *virtual cutoff*. Outside of the plasma, E_b coincides with the reflected wave, while inside the plasma it represents a virtual field that is how the reflected wave would appear to an observer in free space. The position of the virtual cutoff

(r_{vc}) is determined by the phase group delay and is therefore given by

$$r_{vc} = r_0 - c \int_{r_0}^{r_c} \frac{dr}{v_G}, \qquad (7.90)$$

where v_G is the group velocity (7.27). Since the plasma refractive is less than one, $r_{vc} < r_c$. In ionospheric studies, $r_0 - r_{vc}$ is called the *equivalent height* [2] or *virtual height* [39].

As described before, the interference of the spectral components of the reflected wave results in a chaotic pattern in free space, the only place where reflectometry measurements are possible. This is demonstrated in Fig. 7.12(a), which shows a strong modulation of the backward field at the plasma boundary. Furthermore, the phase fluctuation $\tilde{\phi}$ of E_b is completely different from the phase $\tilde{\phi}_{GO}$ (from (7.54)) of geometrical optics (Fig. 7.13(a)).

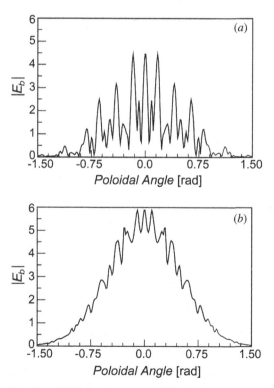

Figure 7.12. Profile of $|E_b|$ for the case of Fig. 7.11 at $r = r_0$ (a) and $r = r_{vc}$ (b). (With permission from [38]).

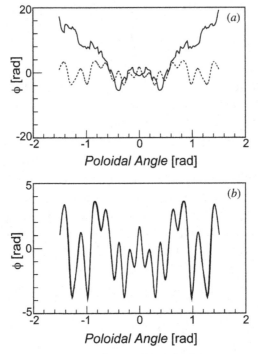

Figure 7.13.   Phase fluctuation $\tilde{\phi}$ of $|E_b|$ (solid line) at $r = r_0$ (a) and $r = r_{vc}$ (b) for the case of Fig. 7.11, versus the phase $\tilde{\phi}_{GO}$ of geometrical optics (dash line). The two curves of plot (b) are identical. (With permission from [38]).

On the contrary at the virtual cutoff, the effect of plasma fluctuations on the amplitude of E_b is small (Fig. 7.12(b)) and, more important, $\tilde{\phi}$ is identical to $\tilde{\phi}_{GO}$ (Fig. 7.13(b)).

Consequently, since most of the contribution to $\tilde{\phi}_{GO}$ comes from a narrow region around the cutoff, the poloidal power spectrum of plasma fluctuations can be obtained from the power spectrum of the phase of E_b at the virtual cutoff, which could be obtained by collecting the scattered waves with a wide aperture antenna and by making an image of the virtual cutoff onto an array of phase sensitive detectors. This is illustrated in Fig. 7.14, where the spectrum of density fluctuations is renormalized for the sake of showing its similarity with the spectrum of the phase of E_b.

Apart from a few cases, such as that of [32], a standard procedure of reflectometry measurements is to identify the spectrum of plasma

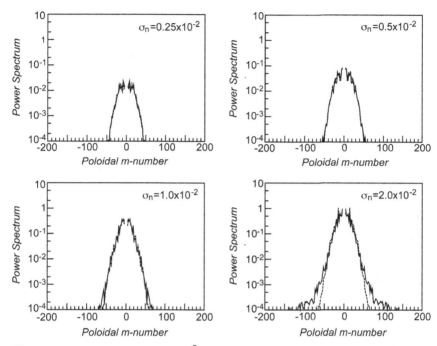

Figure 7.14. Power spectra of $\tilde{\phi}$ at $r = r_{vc}$ (solid line) and of renormalized $\tilde{n}/n$ at $r = r_c$ (dash line) for different values of σ_n. Other parameters are those of Fig. 7.11. Spectra are averaged over twenty realizations of the turbulent field. (With permission from [38]).

turbulence with the spectrum of measured signals. As already stated several times, this leads to erroneous results, as demonstrated by Fig. 7.15, showing the power spectrum of E_b at $r = r_0$ and that of $\tilde{n}/n$ at $r = r_c$. These results demonstrate very clearly that, as plasma fluctuations rise to the level found in fusion experiments, the spectrum of reflected waves in free space becomes considerably broader that the spectrum of plasma turbulence.

Ultimately, the possibility of inferring the spectrum of plasma turbulence from the phase of the backward field breaks down at large levels of plasma fluctuations. This can be explained by the fact that, since each spectral component of the reflected wave with poloidal wave number $k_\theta (= \sigma_\phi \kappa_\theta$ from (7.69)) must originate at a distance $\Delta = L_\varepsilon k_\theta^2 / k_0^2$ from the cutoff, a breakdown of geometrical optics must occur when Δ becomes large than $1/\kappa_r$. From this, since

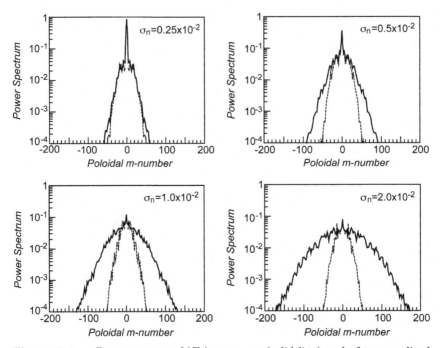

Figure 7.15.   Power spectra of $|E_b|$ at $r = r_0$ (solid line) and of renormalized $\tilde{n}/n$ at $r = r_c$ (dash line) for the fluctuations of Fig. 7.14. Spectra are averaged over twenty realizations of the turbulent field. (With permission from [38]).

$\sigma_\phi^2 \propto \sigma_n^2 k_0^2 L_n / \kappa_r$ (from (7.57)), we get that the critical value of σ_n^2 should scale like

$$\sigma_n^2 \propto \frac{1}{L_n^2 \kappa_\theta^2}. \tag{7.91}$$

A demonstration of the validity of this criterion is illustrated in Fig. 7.16 for the case of $\kappa_r = 1.0\,\mathrm{cm}^{-1}$ and $\kappa_\theta = 1.0\,\mathrm{cm}^{-1}$, showing a dramatic change in the agreement between the measured power spectrum of $\tilde{\phi}$ and that of $\tilde{n}/n$ as σ_n is increase from 1.0×10^{-2} to 2.0×10^{-2}.

The contour plot of $|E_b|$ in Fig. 7.17 shows that this is accompanied by the destruction of the virtual cutoff. Note that the case in Fig. 7.13(b) with the same value of $\sigma_n = 2.0 \times 10^{-2}$ agrees instead with geometrical optics. This is consistent with the criterion (7.91) since the value of κ_θ in Fig. 7.16(b) is twice that in Fig. 7.13(b).

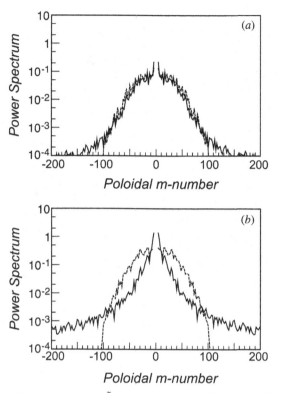

Figure 7.16. Power spectra of $\tilde{\phi}$ at $r = r_{vc}$ (solid line) and of $\tilde{n}/n$ at $r = r_c$ (dash line) for $\kappa_r = 1.0\,\mathrm{cm}^{-1}$, $\kappa_\theta = 1.0\,\mathrm{cm}^{-1}$ and $\sigma_n = 1.0 \times 10^{-2}$ (a) and $\sigma_n = 2.0 \times 10^{-2}$ (b). (With permission from [38]).

As explained in Sec. 7.5, another condition for the validity of (7.54) is for the radial wavelength of fluctuations to be larger than the last lobe of the Airy function, i.e.,

$$\kappa_r < \frac{2k_0}{(k_0 L_\varepsilon)^{1/3}}, \tag{7.92}$$

a condition that is satisfied throughout the present numerical simulation.

In conclusion, the numerical simulation of reflectometry confirms that the characteristics of plasma fluctuations with a short-scale perpendicularly to the direction of the probing wave cannot be determined with reflectometry if the reflected waves are allowed to propagate freely to the point of detection, as in standard reflectometry. The numerical results suggest that, if the amplitude of fluctuations is

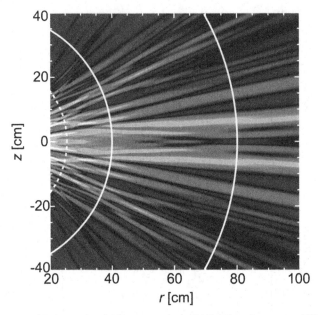

Figure 7.17. (Color online) Contour plot of $|E_b|$ for the case of Fig. 7.16(b) (concentric circles as in Fig. 7.11). (With permission from [38]).

below a threshold that is set by the spectrum of poloidal wave numbers, the local characteristics of density fluctuations can be obtained from the phase of reflected waves when these are collected with a wide aperture antenna, forming an image of the virtual cutoff onto an array of phase sensitive detectors. This method must be considered a technique for the global visualization of turbulent structures in magnetically confined plasmas. Undoubtedly, its practical implementation presents serious difficulties, such as the need for large ports and two-dimenesional arrays of microwave detectors. In any case, reflectometry remains an excellent tool for the study of large-scale plasma fluctuations, such as those driven by magnetohydrodynamic instabilities or Alfven waves [40–43], for which we can ignore the refractive index fluctuation perpendicularly to the propagation of the probing wave.

Bibliography

[1] Ginzburg, V. L., *The Propagation of Electromagnetic Waves in Plasmas*, Pergamon Press, Oxford, 1964.

[2] Budden, K. G., *The Propagation of Radio Waves*, Cambridge University Press, Cambridge, 1985.

[3] Anisimov, A. I., Vinogradov, N. I., Golant, V. E. and Konstantinov, B. P., *Sov. Phys. Tech. Phys.* **5**, 939 (1961).

[4] Anoshkin, V. A., Golant, V. E., Konstantinov, B. P., Poloskin, B. P. and Shcherbinin, O. N., *Sov. Phys. Tech. Phys.* **5**, 1370 (1961).

[5] Doane, J. L., Mazzucato, E. and Schmidt, G. L., *Rev. Sci. Instrum.* **52**, 12 (1981).

[6] Cavallo A. and Cano, R., Technical Report EUR-CEA-FC-1137 (Association EURATOM-CEA sur la Fusion, 1982).

[7] Mazzucato, E., *Bull. Am. Phys. Soc.* **20**, 1241 (1975); Princeton University Plasma Physics Laboratory Report MATT-1151 (1975).

[8] Mazzucato, E., *Phys. Rev. Lett.* **36**, 792 (1976).

[9] Surko C. M. and Slusher, R. E., *Phys. Rev. Lett.* **37**, 1747 (1976).

[10] Mazzucato, E., *Phys. Rev. Lett.* **48**, 1828 (1982).

[11] Brower, D. L., Peebles, W. A., Luhmann, N. C. and Savage, R. L., *Phys. Rev. Lett.* **54**, 689 (1985).

[12] Mazzucato, E., *Rev. Sci. Instrum.* **69**, 2201 (1998).

[13] Fidone, I., Granata, G. and Meyer, L., *Phys. Fluids* **25**, 2249 (1982).

[14] Bornatici, M., Cano, R., De Barbieri, O. and Engelmannn, F., *Nucl. Fusion* **23**, 1153 (1983).

[15] Mazzucato, E., *Phys. Fluids B* **4**, 3460 (1992).

[16] Abramowitz, M. and Stegun, I. A., *Handbook of Mathematical Functions*, Dover, New York, 1965.

[17] Heading, J., *An Introduction to Phase Integral Methods*, Methuen, London, 1962.

[18] Skolnik, E. M., *Introduction to Radar Systems*, McGraw-Hill, New York, 1962.

[19] Doane, J. L., Mazzucato, E. and Schmidt, G. L., *Rev. Sci. Instrum.* **52**, 12 (1981).

[20] Kim, K. W., Doyle, E. J., Peebles, W. A. *et al.*, *Rev. Sci. Instrum.* **66**, 1229 (1995).

[21] Sirinelli, A., Alper, B., Bottereau, C. *et al.*, *Rev. Sci. Instrum.* **81**, 10D939 (2010).

[22] Hanson, G. R., Wilgen, J. B., Bigelow, T. S. *et al.*, *Rev. Sci. Instrum.* **63**, 4658 (1992).

[23] Vershov, V. A., Dreval, V. V. and Soldanov, S. V., in *Proc. 21st EPS Conf. Control. Fusion and Plasma Physics*, Montpellier 18B(III), 1192 (1994).

[24] Heijnen, S. F., Hugenholtz, C. A. J. and Pavlo, P., in *Proc. 18th EPS Conf. Control. Fusion and Plasma Physics*, Berlin 15C(IV), 309 (1991).

[25] Pitteway, M. L., *Proc. Roy. Soc. A* **252**, 556 (1958).

[26] Mazzucato, E. and Nazikian R., *Plasma Phys. Control. Fusion* **33**, 261 (1991).

[27] Ince, I. L., *Ordinary Differential Equations*, Dover, New York, 1956.

[28] Holtzhauer, E., Hirsh, M., Grossmann, T. *et al.*, *Plasma Phys. Control. Fusion* **40**, 1869 (1998).

[29] Hirsh, M., Holtzhauer, E., Baldzuhn, J. *et al.*, *Plasma Phys. Control. Fusion* **43**, 1641 (2001).

[30] Davenport, W. B. and Root, W. L., *An Introduction to the Theory of Random Signals and Noise*, McGraw-Hill, New York, 1957.

[31] Papoulis, A., *Probability, Random Variables and Stochastic Processes*, McGraw-Hill, New York, 1965.

[32] Mazzucato, E. and Nazikian R., *Phys. Rev. Lett.* **71**, 1840 (1993).

[33] Rice, S. O., Bell System Tech. J., **23**, 282 (1944); **24**, 96 (1945) (reprinted in Wax, N., *Selected Papers on Noise and Stochastic Processes*, Dover, New York, 1954).

[34] Beckmann, P. and Spizzichino, A., *The Scattering of Electromagnetic Waves from Rough Surfaces*, Macmillan, New York, 1963.

[35] Mazzucato, E, Batha, S. H., Beer, M. *et al.*, *Phys. Rev. Lett.*, **77**, 3145 (1996).

[36] Mazzucato, E., *Nucl. Fusion* **41**, 203 (2001).

[37] Mazzucato, E., Munsat, T., Park, H. *et al.*, *Phys. Plasmas* **9**, 1955 (2002).

[38] Mazzucato, E., *Plasma Phys. Control. Fusion* **46**, 1271 (2004).

[39] Davies, K., *Ionospheric Radio Waves*, Blaisdell Publishing Co., Waltham MA, 1969.

[40] Nazikian, R., Majeski, R., Mazzucato, E. *et al.*, *Rev. Sci. Instrum.* **68**, 450 (1997).

[41] Nazikian, R., Fu, G.Y., Batha, S. H. *et al.*, *Phys. Rev. Lett.* **78**, 2976 (1997).

[42] Hacquin, S., Alper, B., Sharapov, S. *et al.*, *Nucl. Fusion* **46**, S714 (2006).

[43] Hacquin, S., Sharapov, S., Alper, B. *et al.*, *Plasma Phys. Control. Fusion* **49**, 1371 (2007).

CHAPTER 8

ELECTRON CYCLOTRON WAVES
IN HOT PLASMAS

In the previous chapter we have seen how relativistic effects on both the Hermitian and anti-Hermitian components of the dielectric tensor have profound effects on reflectometry. For this reason, and also in preparation for the subject of the next chapter, here we will review the relativistic theory of wave propagation in a collisionless and uniform plasma imbedded in a constant and homogeneous magnetic field. We will consider waves in the electron cyclotron range of frequencies for which ions can be treated as a static background, providing charge neutrality to the plasma equilibrium.

8.1 Relativistic Theory of Electron Cyclotron Waves

The basic equations in the derivation of the plasma dielectric tensor are the set of equations (2.1)–(2.11) together with the Vlasov equation for the electron distribution function $f(\mathbf{r}, \mathbf{p}, t)$ [1]

$$\frac{\partial f}{\partial t} + \mathbf{v} \cdot \frac{\partial f}{\partial \mathbf{v}} - e\left(\mathbf{E} + \frac{\mathbf{v} \times \mathbf{B}}{c}\right) \cdot \frac{\partial f}{\partial \mathbf{p}} = 0, \qquad (8.1)$$

where $-e$ is the electron charge and the velocity $\mathbf{v}$ and momentum $\mathbf{p}$ are related by

$$\mathbf{p} = \gamma m_e \mathbf{v} \qquad (8.2)$$

with $\gamma = [1 + (p/m_e c)^2]^{1/2} = [1 - (v/c)^2]^{-1/2}$.

As in the case of cold plasma, the problem consists in finding the current density induced by the wave

$$\mathbf{J}(\mathbf{r}, t) = -e \int d\mathbf{p}\, \mathbf{v} f(\mathbf{r}, \mathbf{p}, t). \tag{8.3}$$

We seek solutions in the form

$$
\begin{aligned}
f(\mathbf{r}, \mathbf{p}, t) &= f_0(\mathbf{p}) + f_1(\mathbf{r}, \mathbf{p}, t), \\
\mathbf{E}(\mathbf{r}, t) &= \mathbf{E}_1(\mathbf{r}, t), \\
\mathbf{B}(\mathbf{r}, t) &= \mathbf{B}_0 + \mathbf{B}_1(\mathbf{r}, t),
\end{aligned}
\tag{8.4}
$$

where the subscript 0 indicates the equilibrium quantities and subscript 1 refers to the perturbations induced by the wave. In the following, we will assume that the latter are small and we will neglect terms of higher order. By inserting (8.4) into (8.1) we obtain

$$\mathbf{v} \times \mathbf{B}_0 \cdot \frac{\partial f_0}{\partial \mathbf{p}} = 0, \tag{8.5}$$

$$\frac{\partial f_1}{\partial t} + \mathbf{v} \cdot \frac{\partial f_1}{\partial \mathbf{r}} - \frac{e}{c} \mathbf{v} \times \mathbf{B}_0 \cdot \frac{\partial f_1}{\partial \mathbf{p}} = e \left(\mathbf{E}_1 + \frac{\mathbf{v} \times \mathbf{B}_1}{c} \right) \cdot \frac{\partial f_0}{\partial \mathbf{p}}. \tag{8.6}$$

The most general solution of (8.5) can be written in the form

$$f_0(\mathbf{p}) = f_0(p_\perp, p_\parallel), \tag{8.7}$$

where f_0 is an arbitrary function and subscripts $\perp$ and $\parallel$ indicate the perpendicular and parallel components of $\mathbf{p}$ with respect to $\mathbf{B}_0$. A formal solution of (8.6) can be readily found by noting that the left-hand side term is just the time derivative of f_1 along the zero-order particle trajectory

$$
\begin{aligned}
\frac{d\mathbf{r}}{dt} &= \mathbf{v}, \\
\frac{d\mathbf{p}}{dt} &= -\frac{e}{c} \mathbf{v} \times \mathbf{B}_0,
\end{aligned}
\tag{8.8}
$$

i.e., a *characteristic* of (8.6). Thus, the solution can be written in the form

$$f_1(\mathbf{r}, \mathbf{p}, t) = e \int_{-\infty}^{t} dt' \left(\mathbf{E}(\mathbf{r}', t') + \frac{\mathbf{v}'}{c} \times \mathbf{B}(\mathbf{r}', t') \right) \cdot \frac{\partial f_0(\mathbf{p}')}{\partial \mathbf{p}'}, \quad (8.9)$$

where for simplicity we have dropped the subscript from the perturbed field components. In (8.9), $\mathbf{r}'(t')$ and $\mathbf{p}'(t') = \gamma m_e \mathbf{v}'(t')$ are the solutions of (8.8) satisfying the conditions $\mathbf{r}' = \mathbf{r}$ and $\mathbf{p}' = \mathbf{p}$ at $t' = t$.

It is easy to prove that $p_\perp$, $p_{||}$ and γ are constant along the zero-order particle trajectory. Then, if $(\mathbf{e}_x, \mathbf{e}_y, \mathbf{e}_z)$ is the basis for the system of orthogonal coordinates with $\mathbf{e}_z = \mathbf{B}_0/B_0$, the solution of (8.8) that reaches $\mathbf{v}' = \mathbf{v}$ at $t' = t$ is

$$v'_x = v_x \cos(\omega_c \tau) + v_y \sin(\omega_c \tau),$$
$$v'_y = -v_x \sin(\omega_c \tau) + v_y \cos(\omega_c \tau), \quad (8.10)$$
$$v'_z = v_z = v_{||},$$

where $\omega_c = eB/\gamma m_e c$ is the relativistic electron cyclotron frequency and $\tau = t - t'$. Thus, the zero-order trajectory that reaches $\mathbf{r}' = \mathbf{r}$ at $t' = t$ is

$$x' = x - \frac{v_x}{\omega_c} \sin(\omega_c \tau) - \frac{v_y}{\omega_c}(1 - \cos(\omega_c \tau)),$$
$$y' = y + \frac{v_x}{\omega_c}(1 - \cos(\omega_c \tau)) - \frac{v_y}{\omega_c} \sin(\omega_c \tau), \quad (8.11)$$
$$z' = z - v_z \tau.$$

To proceed, we introduce the space–time Fourier expansion of the electromagnetic field and consider the contribution of a single Fourier component

$$\mathbf{E}(\mathbf{r}, t) = \mathbf{E}(\mathbf{k}, \omega) \exp(i\mathbf{k} \cdot \mathbf{r} - i\omega t),$$
$$\mathbf{B}(\mathbf{r}, t) = \mathbf{B}(\mathbf{k}, \omega) \exp(i\mathbf{k} \cdot \mathbf{r} - i\omega t). \quad (8.12)$$

Since in (8.9) we have tacitly assumed that $f_1|_{t=-\infty} = 0$, we take ω with a small positive imaginary part. Then, from Maxwell equations we get

$$\mathbf{B}(\mathbf{k}, \omega) = \frac{c}{\omega} \mathbf{k} \times \mathbf{E}(\mathbf{k}, \omega), \tag{8.13}$$

which, together with (8.9) and the change of variable $\tau = t - t'$, gives the Fourier amplitude of the distribution function

$$f_1(\mathbf{k}, \omega, \mathbf{p}) = e \int_0^\infty d\tau \exp[i\mathbf{k} \cdot (\mathbf{r}' - \mathbf{r}) + \omega\tau]$$

$$\times \left[\mathbf{E}(\mathbf{k}, \omega) + \frac{\mathbf{v}'}{\omega} \times [\mathbf{k} \times \mathbf{E}(\mathbf{k}, \omega)] \right] \cdot \frac{\partial f_0(\mathbf{p}')}{\partial \mathbf{p}'}. \tag{8.14}$$

By choosing (without loss of generality)

$$\mathbf{k} = k_\perp \mathbf{e}_x + k_{||} \mathbf{e}_z,$$

and the components of the perpendicular velocity in the form

$$v_x = v_\perp \cos(\phi), \quad v_y = v_\perp \sin(\phi),$$

from (8.11), we have

$$\mathbf{k} \cdot (\mathbf{r}' - \mathbf{r}) + \omega\tau = \frac{k_\perp v_\perp}{\omega_c} [\sin(\phi - \phi') - \sin(\phi)] + (\omega - k_{||} v_{||})\tau, \tag{8.15}$$

where $\phi' = \omega_c \tau$. Also, since

$$\frac{\partial f_0}{\partial \mathbf{p}} = \frac{1}{p_\perp} \frac{\partial f_0}{\partial p_\perp} \mathbf{p}_\perp + \frac{\partial f_0}{\partial p_{||}} \mathbf{e}_z,$$

we get

$$\left[\mathbf{E}(\mathbf{k}, \omega) + \frac{\mathbf{v}'}{\omega} \times [\mathbf{k} \times \mathbf{E}(\mathbf{k}, \omega)] \right] \cdot \frac{\partial f_0(\mathbf{p}')}{\partial \mathbf{p}'}$$

$$= (E_x \cos(\phi - \phi') + E_y \sin(\phi - \phi'))$$

$$\times \left[\frac{\partial f_0}{\partial p_\perp} + \frac{k_{||}}{\omega} \left(v_\perp \frac{\partial f_0}{\partial p_{||}} - v_{||} \frac{\partial f_0}{\partial p_\perp} \right) \right]$$

$$+ E_z \left[\frac{\partial f_0}{\partial p_{||}} - \frac{k_\perp}{\omega} \cos(\phi - \phi') \left(v_\perp \frac{\partial f_0}{\partial p_{||}} - v_{||} \frac{\partial f_0}{\partial p_\perp} \right) \right]. \tag{8.16}$$

Then, using the Bessel identity [2]

$$e^{i\rho \sin\theta} = \sum_{n=-\infty}^{\infty} J_n(\rho)e^{in\theta},$$

where $J_n(\rho)$ is the Bessel function of the first kind, the exponential in the integrand of (8.14) becomes

$$e^{i[\mathbf{k}\cdot(\mathbf{r}'-\mathbf{r})+\omega\tau]} = e^{i\rho[\sin(\phi-\phi')-\sin(\phi)]+i(\omega-k_{||}v_{||})\tau}$$

$$= \sum_{m,n=-\infty}^{\infty} J_m(\rho)J_n(\rho)e^{i(n-m)\phi}e^{i(\omega-n\omega_c-k_{||}v_{||})\tau}, \quad (8.17)$$

where $\rho = k_\perp v_\perp/\omega_c$ is equal to 2π times the ratio of the relativistic Larmor radius to the perpendicular wavelength. Inserting (8.16) and (8.17) into (8.14), we find expressions of the type

$$\sum_{n=-\infty}^{\infty} J_n(\rho)e^{i(n-m)\phi}e^{-in\omega_c\tau} \begin{Bmatrix} \cos(\phi - \omega_c\tau) \\ \sin(\phi - \omega_c\tau) \end{Bmatrix}. \quad (8.18)$$

For the first of these we have

$$\sum_{n=-\infty}^{\infty} J_n(\rho)e^{i(n-m)\phi}e^{-in\omega_c\tau}\cos(\phi - \omega_c\tau)$$

$$= \sum_{n=-\infty}^{\infty} \frac{J_n(\rho)}{2}[e^{i(n+1-m)\phi-i(n+1)\omega_c\tau} + e^{i(n-1-m)\phi-i(n-1)\omega_c\tau}]$$

$$= \sum_{q=-\infty}^{\infty} \left[\frac{J_{q-1}(\rho) + J_{q+1}(\rho)}{2}\right]e^{i(q-m)\phi-iq\omega_c\tau}. \quad (8.19)$$

Similarly, for the second expression in (8.18) we have

$$\sum_{n=-\infty}^{\infty} J_n(\rho)e^{i(n-m)\phi}e^{-in\omega_c\tau}\sin(\phi - \omega_c\tau)$$

$$= \sum_{q=-\infty}^{\infty} -i\left[\frac{J_{q-1}(\rho) - J_{q+1}(\rho)}{2}\right]e^{i(q-m)\phi-iq\omega_c\tau}. \quad (8.20)$$

Both expressions can be simplified by using the Bessel identities [2]

$$J_{n-1}(\rho) + J_{n+1}(\rho) = \frac{2n}{\rho} J_n(\rho),$$

$$J_{n-1}(\rho) - J_{n+1}(\rho) = 2J_n'(\rho). \tag{8.21}$$

Assuming, as mentioned above, that ω has a positive imaginary part, the integral over τ in (8.14) can be easily done with the result

$$f_1(\mathbf{k}, \omega, \mathbf{p}) = in_0 e \sum_{m,n=\infty}^{\infty} \frac{J_m(\rho) e^{i(n-m)\phi}}{\omega - n\omega_c - k_{||} v_{||}}$$

$$\times \left\{ \left[E_x \frac{n}{\rho} J_n(\rho) - iE_y J_n'(\rho) \right] F_\perp + E_z J_n(\rho) F_{||} \right\}, \tag{8.22}$$

where n_0 is the equilibrium density, $f_0(\mathbf{p})$ is renormalized such that $\int d\mathbf{p} f_0(\mathbf{p}) = 1$ and

$$F_\perp = \frac{\partial f_0}{\partial p_\perp} + \frac{k_{||} v_\perp}{\omega} \left(\frac{\partial f_0}{\partial p_{||}} - \frac{p_{||}}{p_\perp} \frac{\partial f_0}{\partial p_\perp} \right),$$

$$F_{||} = \frac{\partial f_0}{\partial p_{||}} - n\frac{\omega_c}{\omega} \left(\frac{\partial f_0}{\partial p_{||}} - \frac{p_{||}}{p_\perp} \frac{\partial f_0}{\partial p_\perp} \right). \tag{8.23}$$

Note that the terms inside the brackets become zero for an isotropic electron distribution.

Inserting (8.22) into (8.3) we obtain the Fourier amplitude of the current density. In cylindrical geometry $(p_\perp, \phi, p_{||})$, the integral over ϕ is easily done using the equations

$$\sum_{m=-\infty}^{\infty} J_m(\rho) \int_0^{2\pi} d\phi e^{i(n-m)\phi} \left\{ \begin{matrix} \sin(\phi) \\ \cos(\phi) \\ 1 \end{matrix} \right\} = 2\pi \left\{ \begin{matrix} iJ_n'(\rho) \\ \frac{n}{\rho} J_n(\rho) \\ J_n(\rho) \end{matrix} \right\} \tag{8.24}$$

and we obtain

$$\mathbf{j}(\mathbf{k}, \omega) = -i\frac{n_0 e^2}{m_e} \sum_{n=-\infty}^{\infty} \int \frac{d\mathbf{p}}{\gamma} \frac{\mathbf{S}^{(n)} \cdot \mathbf{E}(\mathbf{k}, \omega)}{\omega - n\omega_c - k_{||} v_{||}}, \tag{8.25}$$

where $\mathbf{S}^{(n)}$ is the tensor

$$
\mathbf{S}^{(n)} =
\begin{pmatrix}
p_\perp \left(\dfrac{nJ_n}{\rho} \right)^2 F_\perp & -ip_\perp \dfrac{nJ_nJ_n'}{\rho} F_\perp & p_\perp \dfrac{nJ_n^2}{\rho} F_\| \\[3mm]
ip_\perp \dfrac{nJ_nJ_n'}{\rho} F_\perp & p_\perp (J_n')^2 F_\perp & ip_\perp J_nJ_n' F_\| \\[3mm]
p_\| \dfrac{nJ_n^2}{\rho} F_\perp & -ip_\perp J_nJ_n' F_\perp & p_\| J_n^2 F_\|
\end{pmatrix} .
\tag{8.26}
$$

This tensor can be cast in a more symmetric form by writing $F_\|$ as

$$
F_\| = \frac{\omega - n\omega_c + k_\| v_\|}{\omega} \left(\frac{\partial f_0}{\partial p_\|} - \frac{p_\|}{p_\perp} \frac{\partial f_0}{\partial p_\perp} \right) + \frac{p_\|}{p_\perp} F_\perp ,
\tag{8.27}
$$

so that, since

$$
\sum_{n=-\infty}^{\infty} nJ_n^2 = 0, \qquad \sum_{n=-\infty}^{\infty} J_nJ_n' = 0,
$$

it becomes

$$
\mathbf{S}^{(n)} =
\begin{pmatrix}
p_\perp \left(\dfrac{nJ_n}{\rho} \right)^2 F_\perp & -ip_\perp \dfrac{nJ_nJ_n'}{\rho} F_\perp & p_\| \dfrac{nJ_n^2}{\rho} F_\perp \\[3mm]
ip_\perp \dfrac{nJ_nJ_n'}{\rho} F_\perp & p_\perp (J_n')^2 F_\perp & ip_\| J_nJ_n' F_\perp \\[3mm]
p_\| \dfrac{nJ_n^2}{\rho} F_\perp & -ip_\| J_nJ_n' F_\perp & p_\| J_n^2 F_\|
\end{pmatrix} .
\tag{8.28}
$$

Note that this tensor satisfies the Onsager symmetry relation (2.26). Finally, from (2.22) we get the dielectric tensor

$$
\boldsymbol{\varepsilon}(\mathbf{k}, \omega) = \mathbf{I} + \frac{\omega_p^2}{\omega} \sum_{n=-\infty}^{\infty} \int \frac{d\mathbf{p}}{\gamma} \frac{\mathbf{S}^{(n)}}{\omega - n\omega_c - k_\| v_\|} .
\tag{8.29}
$$

For an isotropic electron distribution $f_0 = f_0(p)$, the dielectric tensor takes the form

$$
\boldsymbol{\varepsilon}(\mathbf{k}, \omega) = \mathbf{I} + \frac{\omega_p^2}{\omega} \sum_{n=-\infty}^{\infty} \int \frac{d\mathbf{p}}{\gamma} \frac{1}{p} \frac{df_0}{dp} \frac{\mathbf{T}^{(n)}}{\omega - n\omega_c - k_\| v_\|} ,
\tag{8.30}
$$

where the tensor $\mathbf{T}^{(n)}$ (first derived in [3]) is given by

$$\mathbf{T}^{(n)} = \begin{pmatrix} \left(p_\perp \dfrac{nJ_n}{\rho}\right)^2 & -ip_\perp^2 \dfrac{nJ_nJ_n'}{\rho} & p_\perp p_{||} \dfrac{nJ_n^2}{\rho} \\[2ex] ip_\perp^2 \dfrac{nJ_nJ_n'}{\rho} & (p_\perp J_n')^2 & ip_\perp p_{||} J_nJ_n' \\[2ex] p_\perp p_{||} \dfrac{nJ_n^2}{\rho} & -ip_\perp p_{||} J_nJ_n' & (p_\perp J_n)^2 \end{pmatrix}. \tag{8.31}$$

8.2 Dispersion Relation

The general form of the dispersion relation is obtained from (2.31) as

$$D \equiv |\mathbf{NN} - N^2\mathbf{I} + \boldsymbol{\varepsilon}| = 0$$

with $\mathbf{N}$ the refractive index $\mathbf{k}c/\omega = (k_\perp \mathbf{e}_x + k_{||}\mathbf{e}_z)c/\omega = N_\perp \mathbf{e}_x + N_{||}\mathbf{e}_z$. For a dielectric tensor satisfying the Onsager symmetry relation, we get

$$\varepsilon_{xx}N_\perp^4 + 2\varepsilon_{xz}N_{||}N_\perp^3$$
$$+ [N_{||}^2(\varepsilon_{xx} + \varepsilon_{zz}) - \varepsilon_{xx}(\varepsilon_{yy} + \varepsilon_{zz}) + \varepsilon_{xz}^2 - \varepsilon_{xy}^2]N_\perp^2$$
$$+ 2N_{||}(N_{||}^2\varepsilon_{xz} - \varepsilon_{xz}\varepsilon_{yy} + \varepsilon_{xy}\varepsilon_{yz})N_\perp$$
$$+ N_{||}^2(\varepsilon_{xz}^2 - \varepsilon_{yz}^2) + N_{||}^2\varepsilon_{zz}(N_{||}^2 - \varepsilon_{xx} - \varepsilon_{yy})$$
$$+ \varepsilon_{xx}\varepsilon_{yz}^2 - \varepsilon_{yy}\varepsilon_{xz}^2 + 2\varepsilon_{xy}\varepsilon_{xz}\varepsilon_{yz} + \varepsilon_{zz}(\varepsilon_{xy}^2 + \varepsilon_{xx}\varepsilon_{yy}) = 0. \tag{8.32}$$

To proceed, it is convenient to split $\boldsymbol{\varepsilon}$ into Hermitian and anti-Hermitian components, namely

$$\varepsilon_{ij} = \varepsilon'_{ij} + i\varepsilon''_{ij}, \tag{8.33}$$

with

$$\varepsilon'_{ij} = (\varepsilon_{ij} + \varepsilon_{ji}^*)/2, \quad \varepsilon''_{ij} = (\varepsilon_{ij} - \varepsilon_{ji}^*)/2i, \tag{8.34}$$

where the asterisk indicates the complex conjugate.

The main difficulty that one encounters in calculating the dielectric tensor comes from the singular resonance denominator in (8.29).

This can be overcome by adding a small imaginary part to the frequency (as in the previous section), which is then brought down to zero after integration, with the result that by using the Plemelj formula [4] we get

$$\frac{1}{\omega - n\omega_c - k_{||}v_{||}} \rightarrow P\left[\frac{1}{\omega - n\omega_c - k_{||}v_{||}}\right] - i\pi\delta(\omega - n\omega_c - k_{||}v_{||}),$$

(8.35)

where we have assumed $k_{||}$ to be real and P indicates that the Cauchy principal value must be used in the integrand of (8.29).

At this point the algebra becomes quite heavy and beyond the scope of this book, and therefore the interested reader is referred to the pertinent literature [5–9]. Here we will only present the results of [10, 11] for the case of an isotropic electron distribution, which are given in a relatively simple and very useful form. For the anti-Hermitian part, we have (with the axes relabeled from (x, y, z) to $(1, 2, 3)$)

$$\varepsilon''_{11} = a_{11} + N_\perp^2(b_{11} + c_{11}) + N_\perp^4(d_{11} + f_{11} + g_{11}),$$

$$\varepsilon''_{12} = -i[a_{11} + N_\perp^2(2b_{11} + c_{11}) + N_\perp^4(3d_{11} + 3f_{11}/2 + g_{11}),$$

$$\varepsilon''_{22} = a_{11} + N_\perp^2(3b_{11} + c_{11}) + N_\perp^4(37d_{11}/5 + 2f_{11}/2 + g_{11}),$$

$$\varepsilon''_{13} = N_\perp[a_{13} + N_\perp^2(b_{13} + c_{13}) + N_\perp^4(d_{13} + f_{13} + g_{13})],$$

$$\varepsilon''_{23} = iN_\perp[a_{13} + N_\perp^2(2b_{13} + c_{13}) + N_\perp^4(3d_{13} + 3f_{13}/2 + g_{13})],$$

$$\varepsilon''_{33} = N_\perp^2[a_{33} + N_\perp^2(b_{33} + c_{33}) + N_\perp^4(d_{33} + f_{33} + g_{33})].$$

(8.36)

These expressions contain the contribution of the first (a_{ij}, b_{ij}, d_{ij}), second (c_{ij}, f_{ij}) and third (g_{ij}) harmonic of the electron cyclotron frequency and are valid whenever the effect of the forth harmonic is negligible, i.e., for $T_e \leq 25$ keV when $\omega < 2\omega_c$.

The coefficients in (8.36) are given by

$$a_{11} = aRS \sum_{s=\pm 1}\left(1 + \frac{s}{\xi}\right)F(p_s),$$

$$a_{13} = aRS\left(\frac{\omega}{\omega_c}\right) \sum_{s=\pm 1}\left(p_s + \frac{s(p_s + 2sq)}{\xi} + \frac{3sq}{\xi^2}\right)F(p_s),$$

$$a_{33} = aRS \left(\frac{\omega}{\omega_c}\right)^2 \sum_{s=\pm 1} \left(p_s^2 + \frac{sp_s(p_s + 4sq)}{\xi} + \frac{6sq(p_s + sq)}{\xi^2} \right.$$

$$\left. + \frac{12sq^2}{\xi^3} \right) F(p_s),$$

$$b_{11} = aRS \left(\frac{R}{N_{\parallel}\mu}\right) \left(\frac{\omega}{\omega_c}\right)^2 \sum_{s=\pm 1} s \left(1 + \frac{3s}{\xi} + \frac{3}{\xi^2} \right) F(p_s),$$

$$b_{13} = aRS \left(\frac{R}{N_{\parallel}\mu}\right) \left(\frac{\omega}{\omega_c}\right)^3 \sum_{s=\pm 1} s \left(p_s + \frac{3s(p_s + sq)}{\xi} + \frac{3s(p_s + 4sq)}{\xi^2} \right.$$

$$\left. + \frac{15q}{\xi^3} \right) F(p_s),$$

$$b_{33} = aRS \left(\frac{R}{N_{\parallel}\mu}\right) \left(\frac{\omega}{\omega_c}\right)^4$$

$$\times \sum_{s=\pm 1} s \left(p_s^2 + \frac{3sp_s(p_s + 2sq)}{\xi} + \frac{3(p_s^2 + 8sqp_s + 4q^2)}{\xi^2} \right.$$

$$\left. + \frac{30q(p_s + 2sq)}{\xi^3} + \frac{90q^2}{\xi^4} \right) F(p_s),$$

$$d_{11} = aRS \left(\frac{R}{N_{\parallel}\mu}\right)^2 \frac{5}{8} \left(\frac{\omega}{\omega_c}\right)^4 \sum_{s=\pm 1} \left(1 + \frac{6s}{\xi} + \frac{15}{\xi^2} + \frac{15s}{\xi^3} \right) F(p_s),$$

$$d_{13} = aRS \left(\frac{R}{N_{\parallel}\mu}\right)^2 \frac{5}{8} \left(\frac{\omega}{\omega_c}\right)^5 \sum_{s=\pm 1} \left[p_s \left(1 + \frac{6s}{\xi} + \frac{15}{\xi^2} + \frac{15s}{\xi^3} \right) \right.$$

$$\left. + 2\frac{q}{\xi} \left(2 + \frac{15s}{\xi} + \frac{45}{\xi^2} + \frac{105s}{2\xi^3} \right) \right] F(p_s),$$

$$d_{33} = aRS \left(\frac{R}{N_{\parallel}\mu}\right)^2 \frac{5}{8} \left(\frac{\omega}{\omega_c}\right)^6 \sum_{s=\pm 1} \left[p_s^2 \left(1 + \frac{6s}{\xi} + \frac{15}{\xi^2} + \frac{15s}{\xi^3} \right) \right.$$

$$+ 4\frac{qp_s}{\xi} \left(2 + \frac{15s}{\xi} + \frac{45}{\xi^2} + \frac{105s}{2\xi^3} \right)$$

$$\left. + 10\left(\frac{q}{\xi}\right)^2 \left(2 + \frac{15s}{\xi} + \frac{63}{\xi^2} + \frac{84s}{2\xi^3} \right) \right],$$

$$c_{ij} = -4b_{ij}(\omega_c \Rightarrow 2\omega_c), \quad f_{ij} = -128d_{ij}(\omega_c \Rightarrow 2\omega_c)/5,$$

$$g_{ij} = 243d_{ij}(\omega_c \Rightarrow 2\omega_c)/5,$$

where the symbol $b_{ij}(\omega_c \Rightarrow 2\omega_c)$ indicates that ω_c must be replaced by $2\omega_c$ in b_{ij}. The various quantities in these expressions are:

$a = \pi(\omega_p/\omega)^2/4N_{||}^2 K_2(\mu)\exp(\mu)$, $\mu = m_e c^2/T_e$,
$K_2(\mu)$ is the MacDonald function [2], $q = R/(1 - N_{||}^2)$,
$\xi = \mu N_{||} R/(1 - N_{||}^2)$, $R = [(\omega_c/\omega)^2 - 1 + N_{||}^2]^{1/2}$,
S is a step function with $S = 1$ for $R^2 \geq 0$ and $S = 0$ for $R^2 < 0$,
$F(p_s) = \exp[-\mu(\gamma_s - 1)]$, $\gamma_s = (1 - p_s^2)^{1/2}$, with p_s a solution of
$(1 - p_s^2)^{1/2} - \omega_c/\omega = N_{||}p_s$, i.e., $p_s = [N_{||}(\omega_c/\omega) + sR]/(1 - N_{||}^2)$.

For the Hermitian part of ε_{ij} we have

$$\varepsilon'_{11} = 1 + \sum_{s=\pm 1} [A_{11,s} + N_\perp^2(B_{11,s} + C_{11,s})],$$

$$\varepsilon'_{12} = -i \sum_{s=\pm 1} s[A_{11,s} + N_\perp^2(2B_{11,s} + C_{11,s})],$$

$$\varepsilon'_{22} = 1 + \sum_{s=\pm 1} [A_{11,s} + N_\perp^2(3B_{11,s} + C_{11,s})] - 4N_\perp^2 B_{11,s},$$

$$\varepsilon'_{13} = \sum_{s=\pm 1} sN_\perp[A_{13,s} + N_\perp^2(B_{13,s} + C_{13,s})],$$

$$\varepsilon'_{23} = i \sum_{s=\pm 1} N_\perp[A_{13,s} + N_\perp^2(2B_{13,s} + C_{13,s})]$$
$$- i2N_\perp A_{13,0} - (3/2)N_\perp^4 B_{33,0},$$

$$\varepsilon'_{33} = \varepsilon_{330} + N_\perp^2 \sum_{s=\pm 1} [A_{33,s} + N_\perp^2(B_{33,s} + C_{33,s})]$$
$$- 2N_\perp^2 A_{33,0} - (3/2)N_\perp^4 B_{33,0},$$

with

$$\varepsilon_{330} = 1 + \frac{\mu^2(\omega_p^2/\omega^2)}{2K_2(\mu)} \int_{-\infty}^{\infty} du\, u^2 E_i(x_0)\exp(-\mu N_{||}u),$$

$$A_{ij,\sigma} = -\frac{\omega_p^2/\omega^2}{8K_2(\mu)} \int_{-\infty}^{\infty} du \left(\frac{\omega}{\omega_c}u\right)^\delta \Lambda_\sigma(u),$$

$$B_{ij,\sigma} = \frac{\omega_p^2/\omega_c^2}{32\mu^2 K_2(\mu)} \int_{-\infty}^{\infty} du \left(\frac{\omega}{\omega_c}u\right)^{\delta} \tilde{\Lambda}_{\sigma}(u),$$

$$C_{ij,\sigma} = -4B_{ij,u}(\omega/\omega_c \Rightarrow \omega/2\omega_c), \quad \sigma = 0, \pm 1, \quad \delta = \delta_{i3} + \delta_{j3},$$

where

$$\delta_{\alpha 3} = 0 \text{ for } \alpha \neq 3 \quad \text{and} \quad \delta_{\alpha 3} = 1 \text{ for } \alpha = 3,$$

$$\Lambda_{\sigma} = \{1 + (2\mu\gamma_{\|} + x_{\sigma})[1 - x_{\sigma}E_i(x_{\sigma})\exp(-x_{\sigma})]\}\exp(-\mu\gamma_{\|}),$$

$$\tilde{\Lambda}_{\sigma} = \{6 + 2(4\mu\gamma_{\|} + x_{\sigma}) + (2\mu\gamma_{\|} + x_{\sigma})^2,$$

$$\times [1 - x_{\sigma}(1 - x_{\sigma}E_i(x_{\sigma})\exp(-x_{\sigma}))]\}\exp(-\mu\gamma_{\|}),$$

$$x_{\sigma} = -\mu(\gamma_{\|} - \sigma\omega_c/\omega - N_{\|}u), \quad \gamma_{\|} = (1+u^2)^{1/2},$$

$$E_i(x) \equiv \text{exponential integral}.$$

This is the relativistic dielectric tensor that was used for obtaining the results in Fig. 7.4.

8.3   Wave Cutoff in Hot Plasmas

In the cold plasma theory of wave propagation, the cutoff conditions are given by (2.45), which can be written in the compact form

$$\omega_p^2/\omega^2 + \alpha\omega_c/\omega = 1, \tag{8.37}$$

where $\alpha = 0, 1, -1$ for the O, R and L cutoffs, respectively. In the previous chapter, we stated that relativistic effects modify the cutoff conditions, and for the temperatures of interest for a fusion reactor the change can be taken into accounted by simply replacing the electron mass with

$$m = m_e(1 + 5/\mu)^{1/2}, \tag{8.38}$$

where m_e is the rest mass. This can be derived by expanding the Hermitian component of the dielectric tensor in powers of $1/\mu$ near the cutoff where $N_{\perp} \approx 0$, and thus by neglecting finite Larmor radius effects. To first order in $1/\mu$, the only components that survive are [12]

$$\varepsilon'_{11} = \varepsilon'_{22} \approx 1 - \frac{\omega_p^2/\omega^2}{1 - \omega_c^2/\omega^2} \left(1 - \frac{5}{2\mu} \frac{1 + \omega_c^2/\omega^2}{1 - \omega_c^2/\omega^2} \right),$$

$$\varepsilon'_{12} = -\varepsilon'_{21} \approx i \frac{\omega_p^2 \omega_c/\omega^3}{1 - \omega_c^2/\omega^2} \left(1 - \frac{5}{\mu} \frac{1}{1 - \omega_c^2/\omega^2} \right), \qquad (8.39)$$

$$\varepsilon'_{33} \approx 1 - \frac{\omega_p^2}{\omega^2} \left(1 - \frac{5}{2\mu} \right).$$

The justification of (8.38) becomes clear by noting that when $1/\mu \ll 1$, these components of the dielectric tensor can be obtained from those of the cold plasma approximation (2.39) by replacing the electron mass m_e with m. This is a very useful simplification since it is valid for electron temperatures of up to 25 keV ($1/\mu = 0.05$).

From (8.37) and (8.38), then the change Δn_c in the cold cutoff density n_c is

$$\frac{\Delta n_c}{n_c} = \frac{(1 + 5/\mu)^{1/2} - 1}{1 - \alpha \omega_c/\omega}, \qquad (8.40)$$

showing that the largest correction is for the R cutoff.

The use of (8.38) gives not only a good estimate of the cutoff in plasmas with electron temperatures of interest for a fusion reactor, but it provides also a good description of wave propagation in conditions of negligible absorption. This is demonstrated in Fig. 8.1, where the refractive index from the cold plasma approximation using the corrected electron mass is compared with the refractive index from the relativistic approximation for $T_e = 15$ keV and $\omega_c/\omega = 0.8$. The same is demonstrated in Fig. 8.2, showing the radial profile of $N_\perp$ for $\omega/2\pi = 160$ GHz on the equatorial plane of a tokamak plasma, where at the radial location ($R = 2.65$ m) of the magnetic axis, $T_e = 10$ keV, $n_e = 8 \times 10^{19}$ m^{-3} and $\omega_c/\omega = 0.8$. In both cases, the mass correction (8.38) makes the results of the cold approximation in excellent agreement with those of the relativistic approximation.

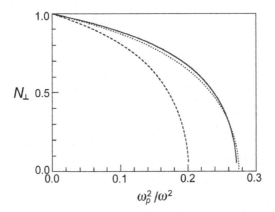

Figure 8.1. X-mode refractive index versus ω_p^2/ω^2 for perpendicular propagation, $T_e = 15\,\text{keV}$ and $\omega_c/\omega = 0.8$. Continuous line is the relativistic approximation; dashed line is the cold plasma approximation; dotted line is the same as the dashed line using the mass correction (8.38). (With permission from [12]).

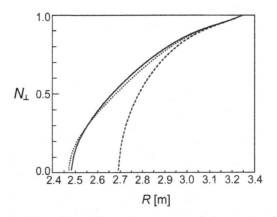

Figure 8.2. Radial profile of $N_\perp$ on the equatorial plane of a tokamak plasma for the X-mode with $\omega/2\pi = 160\,\text{GHz}$. The magnetic axis is at $2.65\,\text{m}$, where $T_e = 10\,\text{keV}$ and $\omega_c/\omega = 0.8$. Lines are labeled as in Fig. 8.1. (With permission from [12]).

Bibliography

[1] Klimontovich, S. R., *The Statistical Theory on Non-equilibrium Processes in a Plasma*, MIT Press, Cambridge, 1967.

[2] Abramowitz, M. and Stegun, I. A., *Handbook of Mathematical Functions*, Dover, New York, 1965.

[3] Trubnikov, B. A., *Plasma Physics and the Problem of Controlled Thermonu-clear Reactions*, Edited by Leontovich, M. A., Vol. III, Pergamon Press, New York, 1958.

[4] Stix, T. H., *Waves in Plasmas*, American Institute of Physics, New York, 1992.

[5] Fidone, I, Granata, G. and Meyer, R. L., *Phys. Fluids* **25**, 2249 (1982).

[6] Airoldi, A. C. and Orefice, A., *J. Plasma Phys.* **27**, 515 (1982).

[7] Bornatici, M., Cano, R., De Barbieri, O. and Engelmann, F., *Nucl. Fusion* **23**, 1153 (1983).

[8] Bornatici, M. and Ruffina, U., *Nuovo Cimento D* **6**, 231 (1985).

[9] Granata, G. and Fidone, I., *J. Plasma Phys.* **45**, 361 (1991).

[10] Fidone, I., Giruzzi, G., Krivenski, V. and Ziebell, L. F., *Nucl. Fusion* **26**, 1537 (1986).

[11] Mazzucato, E., Fidone, I. and Granata, G., *Phys. Fluids* **20**, 3545 (1987).

[12] Mazzucato, E., *Phys. Fluids B* **4**, 3460 (1992).

CHAPTER 9

ELECTRON CYCLOTRON EMISSION

The measurement of plasma emission near harmonics of the electron cyclotron frequency is a ubiquitous diagnostic of electron temperature in present magnetically confined plasma experiments, and will certainly continue to play a major role in the next generation of fusion devices. It is based on the fact that, as we have seen in the previous chapter, a hot plasma imbedded in a magnetic field is a good absorber of electron cyclotron waves and thus, because of the Kirchhoff's law [1], it is also a good emitter of the same waves. Under certain conditions, the intensity of the emission is directly related to the electron temperature by the Planck law. Since the magnetic field of toroidal plasmas is not homogeneous, the electron cyclotron emission (ECE) is a function of position and therefore its measurement can provide a localized value of plasma temperature. In addition, the excellent temporal resolution of the state-of-the-art radiometers allows the detection of temperature fluctuations from both large and small-scale turbulent phenomena. In this chapter, we will review the physics of ECE measurements.

9.1 Radiation Transport

The transport of radiation is best described by its specific intensity (or intensity) I_ω, as defined by

$$d\phi_\omega = I_\omega d\Omega d\omega \cos \alpha d\sigma \tag{9.1}$$

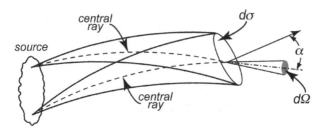

Figure 9.1.   Bundles of rays from a small incoherent source of radiation crossing the area $d\sigma$.

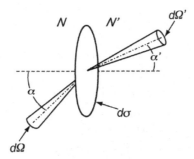

Figure 9.2.   Refraction of a pencil of radiation at the interface between two plasma regions with refractive index N and $N'(>N)$.

where $d\phi_\omega$ is the flux of radiating energy in the frequency range $d\omega$ and solid angle $d\Omega$ when crossing an element of area $d\sigma$ at an angle α with its normal. In the scheme of geometrical optics, this is illustrated in Fig. 9.1 for the case of a small source of incoherent radiation with each of its points sending a bundle of rays through $d\sigma$. The set of central rays fills a cone of solid angle $d\Omega$ with its axis making an angle α with the normal to $d\sigma$.

Let us begin by assuming that the plasma is loss-free and isotropic. Then, at the boundary between two slightly different plasma regions with refractive indexes N and N', as in Fig. 9.2, the power reaching $d\sigma$ must be equal to that leaving the same area, so that

$$I'_\omega d\Omega' d\omega \cos\alpha' d\sigma = I_\omega d\Omega d\omega \cos\alpha d\sigma. \qquad (9.2)$$

From the constancy of $N \sin \alpha$ along the ray (Snell's law of refraction), we have

$$\frac{d\Omega'}{d\Omega} = \frac{\sin \alpha' d\alpha'}{\sin \alpha d\alpha} = \frac{\sin \alpha'}{\sin \alpha} \frac{N \cos \alpha}{N' \cos \alpha'} = \left(\frac{N}{N'}\right)^2 \frac{\cos \alpha}{\cos \alpha'}, \qquad (9.3)$$

so that (9.2) becomes

$$I'_\omega \left(\frac{N}{N'}\right)^2 \cos \alpha d\Omega = I_\omega \cos \alpha d\Omega,$$

from which we readily obtain

$$\frac{d}{ds} \left(\frac{I_\omega}{N^2}\right) = 0, \qquad (9.4)$$

where s is the arc length along the ray.

This equation is not valid in a non-isotropic medium, like a plasma imbedded in a magnetic field, since in this case the rays are tangent to the group velocity — not to the wave vector as in the case of isotropic media. This is clearly seen from the first of the ray equation (3.28), showing that $d\mathbf{r}/ds$ is not parallel to the wave vector $\mathbf{k}$ when $\partial D/\partial \theta \neq 0$ (where θ is the angle between $\mathbf{k}$ and the magnetic field). From this, we get

$$\tan \beta = \frac{1}{k} \frac{\partial D/\partial \theta}{\partial D/\partial k} = \frac{1}{k} \frac{\partial k}{\partial \theta} \qquad (9.5)$$

for the angle β between $\mathbf{k}$ and the ray direction. The result is that the Snell's law is not valid for anisotropic plasmas. In [2], it is shown that in this case (9.3) must be replaced by

$$\frac{d\Omega'}{d\Omega} = \left(\frac{N_r}{N'_r}\right)^2 \frac{\cos \alpha}{\cos \alpha'}, \qquad (9.6)$$

where

$$N_r^2 = N^2 \left| \frac{\sin \theta}{\cos \beta [\partial \cos(\theta - \beta)/\partial \theta]} \right|, \qquad (9.7)$$

which we might call the *ray refractive index*. Thus for anisotropic plasmas, (9.4) must be replaced by

$$\frac{d}{ds}\left(\frac{I_\omega}{N_r^2}\right) = 0. \tag{9.8}$$

To complete the equation of radiation transport we need to add absorption and emission. This can be readily done by assuming that by moving a distance ds the ray suffers the absorption of energy

$$-\alpha_\omega I_\omega ds d\sigma \cos \alpha d\Omega d\omega, \tag{9.9}$$

where $\alpha_\omega(s)$ is the absorption coefficient.

Emission can be added by defining the emission coefficient j_ω as the power generated per unit volume, frequency and solid angle flowing in the ray direction, so that the contribution to (9.8) is

$$j_\omega ds d\sigma \cos \alpha d\Omega d\omega. \tag{9.10}$$

The result is the *equation of radiative transfer*

$$N_r^2 \frac{d}{ds}\left(\frac{I_\omega}{N_r^2}\right) = j_\omega - \alpha_\omega I_\omega. \tag{9.11}$$

9.2 Solution of Radiative Transfer Equation

It is useful to define two quantities. The first is the *source function*

$$S_\omega = \frac{1}{N_r^2}\frac{j_\omega}{\alpha_\omega}.$$

The other is the *optical depth* τ, as defined by

$$d\tau = \alpha_\omega ds. \tag{9.12}$$

The equation of radiative transfer can then be written as

$$\frac{d}{d\tau}\left(\frac{I_\omega}{N_r^2}\right) = S_\omega - \frac{I_\omega}{N_r^2}. \tag{9.13}$$

This equation can be easily integrated by introducing the function $f(\tau) = e^\tau I_\omega(\tau)/N_r^2(\tau)$, satisfying the equation $df/d\tau = e^\tau S_\omega$, from

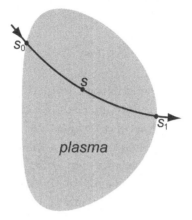

Figure 9.3. A ray passing through the plasma (surrounded by a vacuum region) from entrance at $s = s_0$ to the point of emergence at $s = s_1$ (direction of arc-length coordinate s is the same as that of the ray trajectory).

which we obtain

$$\frac{I_\omega(s)}{N_r^2(s)} = I_\omega(s_0)e^{-\tau} + \int_0^\tau S_\omega(\tau')e^{-(\tau-\tau')}d\tau', \qquad (9.14)$$

where (see Fig. 9.3)

$$\tau(s) = \int_{s_0}^s \alpha_\omega ds, \qquad (9.15)$$

and we have assumed the plasma to be surrounded by a vacuum region, so that $N_r(s_0) = N_r(s_1) = 1$. At the exit point $s = s_1$, we have

$$I_\omega(s_1) = I_\omega(s_0)e^{-\tau_0} + \int_0^{\tau_0} S_\omega(\tau')e^{-\tau'}d\tau', \qquad (9.16)$$

where the optical depth has been redefined as

$$\tau(s) = \int_s^{s_1} \alpha_\omega ds \qquad (9.17)$$

and $\tau_0 = \tau(s_0)$ is the total optical depth. In (9.16), the term $I_\omega(s_0)$ comes from the plasma emission flowing in the ray opposite direction (Fig. 9.3), which is reflected by the wall of the vessel containing the plasma and sent back to $s = s_1$.

If η_w is the wall reflectivity, this term is given by

$$I_\omega(s_0) = \eta_w \int_0^{\tau_0} S_\omega(\tau')e^{-(\tau_0-\tau')}d\tau'. \tag{9.18}$$

For this to give a significant contribution to (9.16), the total optical depth τ_0 must not be too large and η_w must be close to unity. Similarly, the contribution of the second term on the right-hand side of (9.16) can only come from plasma regions with small values of τ.

To complete the radiative transport equation, we must introduce the emission coefficient j_ω for a plasma in thermodynamic equilibrium, which is given by the Kirchhoff's law [1]

$$S_\omega = \frac{1}{N_r^2}\frac{j_\omega}{\alpha_\omega} = \frac{\omega^2 T_e}{8\pi^3 c^2}. \tag{9.19}$$

The right-hand side of (9.19) is the Rayleigh–Jeans blackbody intensity of one mode in an anisotropic medium [2, 3], which is valid when quantum effects are not important. i.e., $\hbar\omega \ll T_e$, a condition that is satisfied for the plasmas on interest in this book.

In conclusion, an observer at $s = s_1$ would detect a blackbody emission with radiation temperature

$$\begin{aligned}T_r &= \eta_w e^{-\tau_0}\int_{s_0}^{s_1} T_e(s')\alpha_\omega(s')e^{-(\tau_0-\tau(s'))}ds' \\ &\quad + \int_{s_0}^{s_1} T_e(s')\alpha_\omega(s')e^{-\tau(s')}ds'.\end{aligned} \tag{9.20}$$

As we shall see in the next section, in the hot plasmas of interest for fusion research $\tau_0 \gg 1$, and thus (9.20) becomes

$$T_r = \int_{s_0}^{s_1} G(s')T_e(s')ds', \tag{9.21}$$

where $G(s') = \alpha_\omega(s')e^{-\tau(s')}$ plays the function of a weighting factor, which because of (9.17) and the assumption $\tau_0 \gg 1$ satisfies the equation

$$\int_{s_0}^{s_1} G(s')ds' = \int_0^{\tau_0} e^{-\tau}d\tau \approx 1.$$

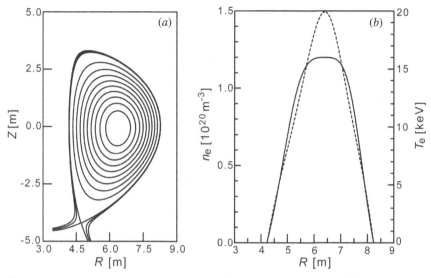

Figure 9.4. Magnetic configuration (a) and radial profiles (b) of electron density (full line) and temperature (dash line).

9.3 Plasma Emission

In this section we discuss how (9.21) can be used together with the theory of electron cyclotron waves for the measurement of electron temperatures. As test-bed, we take the tokamak plasma configuration of Fig. 9.4, which is what is expected for ITER [4]. Again, the reason for choosing a tokamak is that the ECE diagnostic, as many of the techniques described in this book, was developed for this type of magnetic configuration [5–14].

Since the fundamental plasma parameter in the measurement of electron temperatures with ECE is the temperature itself, we will vary the value of T_e keeping everything else constant. Indeed, this is not completely correct since a tokamak magnetic configuration depends on the plasma beta, i.e., on the plasma temperature. However, this will simplify our analysis and provide a clear demonstration of the role played by T_e.

As already mentioned in the introduction, the spatial dependence of the toroidal magnetic field in tokamaks, which approximately varies like $1/R$, leads to the possibility of obtaining spatially

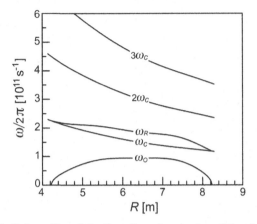

Figure 9.5. Radial profiles of the first three harmonics of the electron cyclotron frequency together with the ordinary (ω_O) and extraordinary (ω_R) cutoff frequencies (with (8.38) correction) for the tokamak of Fig. 9.4 with $B_T = 5.3$ T.

resolved measurements of ECE. Assuming for the plasma of Fig. 9.4 a toroidal magnetic field of 5.3 T at $R = 6.2$ m and a toroidal current of 15 mA — same as in ITER — Fig. 9.5 shows the radial dependence of the first three harmonics of the electron cyclotron frequency (for cold plasma) together with the cutoff frequencies of the ordinary and extraordinary modes (with correction (8.38)).

The use of (9.21) requires knowledge of the absorption coefficient α_ω, which can be obtained from (3.25) as

$$\alpha_\omega = 2\mathbf{k}^i \cdot \frac{\mathbf{v}_G}{v_G}, \qquad (9.22)$$

where $\mathbf{k}^i$ is the imaginary component of the wave vector and $\mathbf{v}_G$ is the group velocity. As often in the experiments, we assume that the observer's line of sight is on the equatorial plane perpendicularly to the magnetic surfaces, so that (9.22) becomes

$$\alpha_\omega = 2k_0 N_\perp^i, \qquad (9.23)$$

where $k_0 = \omega/c$ and $N_\perp^i$ is the imaginary component of the solution of the dispersion relation (8.32). Figure 9.6 shows $N_\perp^i$ as a function of position for the extraordinary mode with $\omega/2\pi = 285$ GHz. Here it is important to note that $N_\perp^i \neq 0$ only for $R < 6.8$ m, which is the region where $\omega < 2\omega_c$. This is consistent with the values of $N_\perp^i$ being

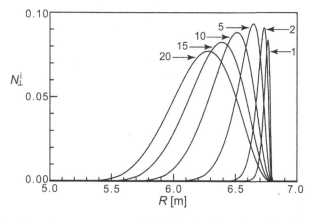

Figure 9.6. Imaginary component of $N_\perp$ for the X-mode with $\omega/2\pi = 285\,\text{GHz}$ on the equatorial plane of the tokamak plasma of Fig. 9.4. Labels are central values of T_e in keV.

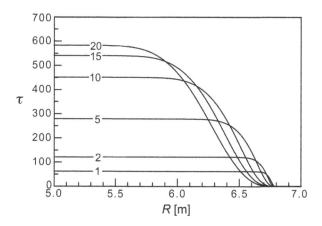

Figure 9.7. Optical depth for the cases in Fig. 9.6.

only the contribution of the second harmonic for which the resonance condition is $\omega = 2\omega_c/\gamma$ (with $k_\parallel = 0$) and thus $\omega < 2\omega_c$.

From the corresponding large values of τ (Fig. 9.7) we conclude that the first term on the right-hand side of (9.20) is totally insignificant, while the second originates in a narrow region near the point where $\omega = 2\omega_c$. This is illustrated in Fig. 9.8, which displays the weighting factor G. In all six cases, the radiation temperature is in very good agreement with the value of T_e near the maximum of the

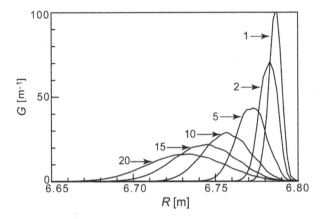

Figure 9.8. G-factor for the cases in Fig. 9.6.

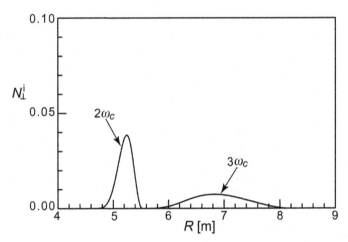

Figure 9.9. $N_\perp^i$ for the X-mode with frequency of 350 GHz and a central temperature of 25 keV.

function G. This shows that even in the case of 20 keV — with its broad $N_\perp^i$ — the ECE measured temperatures would remain quite localized.

However, as the electron temperature increases, this begins to breakdown as illustrated in Fig. 9.9, which shows a case of harmonic overlap for the frequency of 350 GHz and a central temperature of 25 keV. This occurs when the emission from two or more harmonics have the same frequency, so that in Fig. 9.9 we must have

$$\frac{350}{2\pi} = \frac{2\omega_c(r_2)}{\gamma_2} = \frac{3\omega_c(r_3)}{\gamma_3},$$

where r_2 and r_3 are the radii of the second and third harmonic res-
onances with γ_2 and γ_3 the corresponding values of gammas. The
two harmonics can resonate at the same location only if $\gamma_3 = 3\gamma_2/2$,
which makes γ_3 too large for the case of Fig. 9.9. However, emission at
the third harmonic may occur at a different location from that of the
second harmonic, with the result that one resonance can absorb the
emission from the other, as demonstrated in Fig. 9.10 showing that
an observer on the low field side of the plasma would detect the
plasma emission from the outside edge of the plasma rather than
from $R \approx 5.3$.

Finally, since the relative separation of two adjacent harmonics
varies inversely to their number, their overlap becomes more important for high harmonics.

In conclusion, the measurement of ECE is a powerful technique
for the measurement of the electron temperature in magnetically
confined plasmas. However, the probing region is restricted by the
overlap of harmonics that may occur at high plasma temperatures.

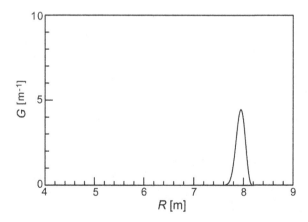

Figure 9.10. *G*-factor for the case of Fig. 9.9.

9.4   Temperature Measurements

The instruments for ECE measurements fall into two classes: quasi-optical systems including the *Fourier-transform spectrometer* [5, 6] and the *grating spectrometer* [7–9], and those employing microwave techniques, such as the *heterodyne receiver* [10–14].

The *Fourier-transform spectrometer*, which is based on the Martin–Puplett version of the Michelson interferometer [15, 16], is schematically illustrated in Fig. 9.11. Here, the ECE from the plasma is first split into two beams that are then recombined after one is reflected from a fast scanning mirror and the other from a fixed mirror. The ECE spectrum is obtained by the Fourier transform of the resulting interferogram with a variable path difference [17]. Main advantages are the possibility of obtaining spectra over several harmonics and a large optical throughput. The disadvantage is that of a modest frequency resolution $(\Delta\nu)$. The latter is determined by the path difference (ΔL) between fixed and scanning mirrors and it is given by $\Delta\nu = c/2\Delta L$, usually $\Delta L = 3-10\,\mathrm{cm}$ and $\Delta\nu = 1.5-5\,\mathrm{GHz}$. The temporal resolution depends on the scanning velocity and is in the range of 2–10 ms. It can be increased by replacing the scanning flat mirror with a rotating mirror containing several helicoidal sectors (Fig. 9.11) [18].

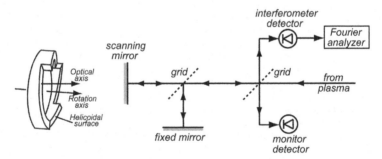

Figure 9.11.   Right: schematic diagram of a Fourier-transform spectrometer based on the Martin–Puplett version of the Michelson interferometer. The grid acts as polarizing semitransparent mirrors. Left: rotating helicoidal mirror that may replace the flat scanning mirror.

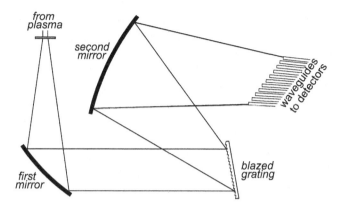

Figure 9.12. Grating spectrometer for ECE measurements.

A *grating spectrometer* for ECE measurements is schematically shown in Fig. 9.12. It works in combination with a system of multiple waveguides, each carrying the output to a detector. The time resolution is very high, of the order of $1\,\mu s$, and the ECE emission can be measured simultaneously at several frequencies with good resolution ($\sim 5 \times 10^{-2}$). Its major drawback is a low throughput.

Two schematic diagrams of a *heterodyne receiver* are shown in Fig. 9.13. The first is a double sideband receiver, where the ECE signal is mixed with a local oscillator (LO) in a nonlinear device (mixer) where it is down converted into a range of intermediate frequency (IF), which is then amplified, filtered and detected. The resulting signal is a measure of the ECE power in two narrow sidebands of the LO frequency. The second diagram is that of a broadband multichannel receiver where again the ECE signal is down converted to an IF range of frequencies and amplified. It is then split into n parts that are filtered by a series of narrowband filters and sent to an array of detectors, providing a measure of the full spectrum of ECE. In both cases, the video stage consists of a square-law detector followed by a low-pass integrating filter. These receivers have a very good frequency resolution ($\sim 10^{-3}$) and an excellent time resolution (less than $1\,\mu s$), limited mainly by the signal-to-noise ratio. In addition, they can use microwave detectors that operates at room temperature, whereas

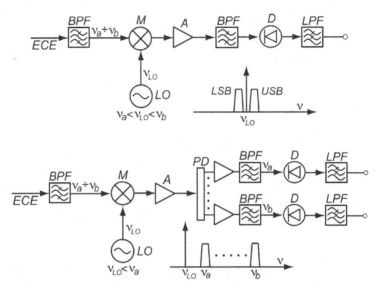

Figure 9.13. Block diagram of a single channel double sideband receiver (top) and of a multichannel receiver (bottom). LO = local oscillator, M = mixer, A = amplifier, BPF = band-pass filter, LPF = low-pass filter, PD = power divider, D = square-law detector. Graphs at the bottom right of each diagram represent the spectral coverage of the ECE measurement.

those of all infrared spectrometers need to be cooled at liquid-helium temperatures. The disadvantage is that very broadband microwave mixers are available only up to frequencies of 250–300 GHz.

9.5 Temperature Fluctuations Measurements

In addition to the measurement of the temporal evolution of the electron temperature in magnetically confined plasmas, the excellent time resolution of modern radiometers in the microwave/millimeter part of the electromagnetic spectrum allows ECE to be used for the detection of turbulent temperature fluctuations with both large and short scales.

The application of ECE to the study of large-scale events, such as those driven by magnetohydrodynamic instabilities, is indeed a straightforward extension of the temperature profile measurements [19, 20]. Here, a recent instrumental development has been the use of 1D linear array of detectors [21–24], where, by making an image of the plasma onto the array and by sweeping the frequency of the local oscillator, one can obtain a 2D image of the poloidal cross-section of tokamak plasmas.

On the contrary, the use of ECE for measuring small short-scale temperature fluctuations, such as those induced by a broadband turbulence, is not a completely trivial matter since what one trys to do is to detect plasma noise using plasma noise, i.e., plasma turbulence with plasma thermal emission.

Before describing this type of measurements, let us discuss the importance of temperature fluctuations. As mentioned already, numerous theories and numerical simulations support the conjecture that the anomalous transport of plasma energy that is observed in magnetized plasmas may arise from some kind of short-scale electrostatic turbulence [25], for which the induced electron energy flux across the magnetic surfaces is

$$Q_e = \frac{3}{2}\overline{\tilde{p}_e \tilde{v}_e} = \frac{3}{2}\frac{n_e \overline{\tilde{T}_e \tilde{E}_\perp} + T_e \overline{\tilde{n}_e \tilde{E}_\perp}}{B}, \qquad (9.24)$$

where the over-bar indicates the average, $\tilde{n}_e$, $\tilde{T}_e$, $\tilde{p}_e$ are fluctuations in electron pressure, density and temperature and $\tilde{E}_\perp$ is the component of the turbulent electric field ($\tilde{\mathbf{E}}$) on the magnetic surface that is perpendicular to the magnetic field $\mathbf{B}$, so that $\tilde{v}_c = \tilde{E}_\perp/B$ is the perpendicular component of the $\tilde{\mathbf{E}} \times \mathbf{B}$ drift velocity to the magnetic surface.

Probe measurements of fluctuating temperature, density and electric potential at the edge of magnetized plasmas indicate that the observed electron transport in tokamaks can indeed be accounted by (9.24). Unfortunately, this type of measurements cannot be performed in the core of hot plasmas since methods for core measurements of the electric potential are not available. Nevertheless, the importance of density and temperature fluctuation measurements is that they can provide, in combination with numerical simulations of plasma turbulence, crucial information on the nature of anomalous transport in magnetized plasmas, which so far has been the major obstacle to the realization of a fusion reactor.

The difficulty in detecting these fluctuations stems from the fact that their relative amplitude is many orders of magnitude smaller than the intrinsic noise of ECE, as it can be seen by considering the quantum expression for the fluctuation in the number of photons per quantum state of a blackbody, given by [1]

$$\frac{\overline{\delta n_s^2}}{\bar{n}_s^2} = 1 + \frac{1}{\bar{n}_s}, \tag{9.25}$$

where $\bar{n}_s$ is the average number of photons [1]

$$\bar{n}_s = \frac{1}{e^{h\nu/T_e} - 1}. \tag{9.26}$$

When $h\nu/T_e \ll 1$, which here is the case of interest, $\bar{n}_s \approx T_e/h\nu$, so that

$$\frac{\overline{\delta n_s^2}}{\bar{n}_s^2} \approx 1, \tag{9.27}$$

from which we get that the noise energy per quantum state is

$$\tilde{E}_s = h\nu(\overline{\delta n_s^2})^{1/2} = T_e. \tag{9.28}$$

To find the energy of a set of N adjacent quantum states, we can take advantage of the fact that, since we are dealing with a frequency band $\Delta\nu$ that is much smaller than ν, we can assume that $\bar{n}_s$ is constant over this band. Thus, for the average energy of N quantum states we have

$$\bar{E}^2 = (h\nu)^2 \overline{\left(\sum_{i=1}^{N} n_i\right)^2} = (h\nu)^2 N^2 \bar{n}_s^2, \tag{9.29}$$

where we have taken into account the statistical independence of quanta. Similarly

$$\overline{(E^2 - \bar{E}^2)} = (h\nu)^2 \overline{\left(\left(\sum_{i=1}^{N} n_i\right)^2 - N^2 \bar{n}_s^2\right)}$$

$$= (h\nu)^2 \overline{\left(\sum_{i=1}^{N} (n_i^2 - \bar{n}_s^2)\right)} = (h\nu)^2 N^2 \overline{\delta n_s^2}. \tag{9.30}$$

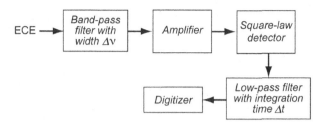

Figure 9.14. Basic components of a plasma radiometer.

From these and (9.27) we have

$$\frac{\delta P_{rms}}{\bar{P}} = 1, \tag{9.31}$$

i.e., the root-mean-square amplitude of received power is 100% of the average power, and thus would mask any small temperature fluctuation induced by plasma turbulence.

These large ECE fluctuations can be partially reduced by using the scheme illustrated in the block diagram of Fig. 9.14, where the ECE signal within a narrow frequency band $\Delta\nu$ is first detected by a square-law detector and is then filtered by a low-pass filter (LPF) with integration time-constant Δt. The function of the latter can be understood as the multiple sampling of the radiation intensity. If we call $\tau_c = 1/\Delta\nu$ the radiation correlation time, the number of independent samples is $N_s = \Delta t/\tau_c$, so that (9.31) becomes

$$\frac{\delta P_{\rm rms}}{\bar{P}} = \frac{1}{\sqrt{N_s}}. \tag{9.32}$$

If we define $\Delta\nu_D = 1/2\Delta t$ as the post-detection bandwidth, this can be cast in the form

$$\frac{\delta P_{\rm rms}}{\bar{P}} = \sqrt{\frac{2\Delta\nu_D}{\Delta\nu}}. \tag{9.33}$$

Since the smoothing of the ECE signal by the LPF must not modify the signal modulation by the plasma temperature fluctuations, $\Delta\nu_D$ must not be smaller than the frequency range of the latter. On the other hand, the value of $\Delta\nu$ is limited by the desired radial resolution, given by $\Delta R/R = \Delta\nu/n\nu_c$, where n is the ECE harmonic number and ν_c is the electron cyclotron frequency. For $\Delta\nu = 1\,{\rm GHz}$ and

$n = 2$, which would give a radial resolution of about $1\,\mathrm{cm}$ in the main core of the plasma of Fig. 9.5, and $\Delta\nu_D = 200\,\mathrm{kHz}$, which is the frequency range of expected plasma temperature fluctuations, from (9.33) we obtain $\delta P_{\mathrm{rms}}/\bar{P} = 2.0 \times 10^{-2}$, which is close to the magnitude of expected turbulent fluctuations.

Even though this is quite a reduction in the radiation noise, the latter is still large enough to prevent a direct measurement of temperature fluctuations with ECE. Fortunately, a further reduction is possible through the use of correlation techniques by exploiting the random nature and the short correlation time of the thermal noise. The methods consists in performing a correlation analysis of two distinct signals, S_1 and S_2, each containing the same information on the plasma temperature fluctuation $\tilde{T}$ together with two uncorrelated components $(\tilde{N}_1, \tilde{N}_2)$ of the thermal noise. Correlation of the two signals over a suitable long time gives

$$\langle S_1 S_2 \rangle = \langle \tilde{T}^2 \rangle + \langle \tilde{N}_1 \tilde{N}_2 \rangle + \langle \tilde{T}\tilde{N}_1 \rangle + \langle \tilde{T}\tilde{N}_2 \rangle \approx \langle \tilde{T}^2 \rangle. \qquad (9.34)$$

If N is the number of averaging samples, from (9.33) and the theory of stationary random processes [26] we get that the temperature measurement sensitivity is given by

$$\frac{\langle \tilde{T}^2 \rangle}{\langle T \rangle^2} = \frac{1}{\sqrt{N}} \frac{2\Delta\nu_D}{\Delta\nu}. \qquad (9.35)$$

For example, the noise in the temperature measurement is reduced by a factor of 30 for $N \sim 10^6$. For the case mentioned above, this would lower the measurement uncertainty to $\sim 0.07\%$.

Since the pioneering works on the measurement of temperature fluctuations with ECE [27, 28], the technique has been applied on several tokamaks [29–32]. In general, the cross-correlation is done in Fourier space by using the Fourier transform of two radiometer signals from stationary plasmas.

Figure 9.15 illustrates two correlation schemes for temperature fluctuations measurements. In the first (top of Fig. 9.15), the common antenna of two radiometers detects the ECE radiation in two disjoint frequency bands ν_1 and ν_2, such that the corresponding emitting plasma layers are separated by a distance shorter than the correlation length of temperature fluctuations but longer than that of the

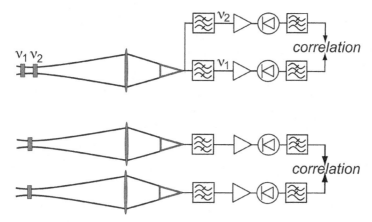

Figure 9.15. Two correlation schemes for temperature fluctuation measurements.

plasma thermal emission. The two frequency bands are then separated by band pass filters and detected by two radiometers, whose signals are Fourier analyzed and cross-correlated giving the spectrum of temperature fluctuations. In principle, by varying the frequency separations of the two bands one could obtain the correlation length of temperature fluctuations perpendicularly to the magnetic field.

The second scheme (bottom of Fig. 9.15) takes advantage of the long wavelength of turbulent fluctuations along the magnetic field lines, which is of the order of the connection length [25], defined as the length that it takes to a field line to complete a full poloidal turn. In tokamaks, this is equal to $2\pi q R$, where q is the magnetic safety factor and R is the plasma major radius. Thus, two radiometers viewing different plasma volumes on the same field line separated by a distance of the order of the radiation pattern of the receiving antenna will detect coherent temperature fluctuations and completely decorrelated thermal noise. The cross-correlation procedure then is the same as in the previous case. Other correlation schemes are described in [33].

The essence of these measurements is the correlation of the ECE emission from two plasma regions containing the same information on temperature fluctuations together with uncorrelated components of the thermal noise. From plasma theory [25] and measurements

of density fluctuations [34, 35], we know that the correlation length
($\delta_\perp$) of small-scale turbulent fluctuations in the plane perpendicular to the magnetic field is of the order of a few ion Larmor radii.
Hence, the correlation analysis must be performed on ECE signals
coming from plasma regions that do not extend across the magnetic
field by more than a few centimeters, otherwise the result of the measurement would just be a meaningless spatial average of temperature
fluctuations — not their average amplitude. The problem stems from
the fact that, as Fig. 9.8 clearly demonstrates, the radial width of
the emission region increases very rapidly with the electron temperature. While this is not a problem for temperature measurements,
it is a serious impediment to temperature fluctuation measurements
in hot plasmas. This issue is rarely mentioned in the literature, the
reason being that this type of measurements have been performed
only in relatively cold (<2 keV) and small ($R < 2$ m) plasmas (the
plasma size being important since the radial width of the weighting
G-factor varies linearly with R). However, an ECE measurement of
temperature fluctuations can still be performed in the outer and cold
region of hot plasmas that recent experiments [36] and theories [37]
indicate being the site of turbulent processes that may affect plasma
conditions in the central region.

In conclusion, ECE fluctuation measurements offer a unique
opportunity for the study of short-scale temperature fluctuations in
relatively cold (T_e less than a few keV) and small plasmas (R less than
1–2 meters). Unfortunately, these measurements will not be possible
in the core of hot and large plasmas of thermonuclear reactors.

Bibliography

[1] Reif, F., *Fundamental of Statistical and Thermal Physics*, McGraw-Hill,
     New York, 1965.
[2] Bekefi, G., *Radiation processes in Plasmas*, Wiley, New York, 1966.
[3] Bornatici, M., Cano, R., De Barbieri, O. and Engelmann, F., *Nucl. Fusion*
     23, 1153 (1983).
[4] ITER Technical Basis, *ITER EDA Documentation Series No.* 24, IAEA,
     Vienna, 2002.
[5] Costley, A. E., Hastie, R. J., Paul, J. W. M. and Chamberlain, J., *Phys.
     Rev. Lett.* **33**, 758 (1974).

[6] Costley, A. E. and TFR Group, *Phys. Rev. Lett.* **38**, 1477 (1977).

[7] Tait, G. D., Stauffer, F. J. and Boyd, D. A., *Phys. Fluids* **24**, 719 (1981).

[8] Fischer, J., Boyd, D. A., Cavallo, A. and Benson J., *Rev. Sci. Instrum.* **54**, 1085 (1983).

[9] Cavallo, A., Cutler, R. C., and McCarthy, M. P., *Rev. Sci. Instrum.* **59**, 889 (1988).

[10] Hosea, J. C., Arunasalam, V. and Cano, R., *Phys. Rev. Lett.* **39**, 408 (1977).

[11] Cano, R., Bagdasarov, A. A., Berlizov, A. B., Gorbunov, E. P. and Notkin, G. E., *Nucl. Fusion* **9**, 1415 (1979).

[12] Efthimion, P. C., Arunasalam, V. and Hosea, J. C., *Phys. Rev. Lett.* **44**, 396 (1980).

[13] Taylor, G. Efthimion, P., McCarthy, M., Arunasalam, V. *et al.*, *Rev. Sci. Instrum.* **55**, 1739 (1984).

[14] Hartfuss, H. J. and Tutter, M., *Rev. Sci. Instrum.* **56**, 1703 (1985).

[15] Martin, D. H. and Paplett, E., *Infrared Phys.* **10**, 105 (1970).

[16] Martin, D. H., in *Infrared and Millimeter Waves*, Edited by Button, K. J., Vol. 6, Chap. 3, Academic Press, New York, 1982.

[17] Chantry, G. W., *Submillimeter Spectroscopy*, Academic Press, London, 1971.

[18] Buratti, P. and Zerbini, M., *Rev. Sci. Instrum.* **66**, 4208 (1995).

[19] Nagayama, Y., Sabbagh, S. A., Manickarn, J., Fredrickson, E. D. *et al.*, *Phys. Rev. Lett.*, **69**, 2376 (1992).

[20] Chang, Z., Park, W., Fredrickson, E. D. and Batha, S. H. *et al.*, *Phys. Rev. Lett.*, **77**, 3553 (1996).

[21] Park, H. K., Mazzucato, E., Munsat, T., Domier, C. W. and Johnson, M., *Rev. Sci. Instrum.* **75**, 3787 (2004).

[22] Park, H. K., Luhmann Jr., N. C., Donné, A. J. H., Classen, I. G. J. *et al.*, *Phys. Rev. Lett.* **96**, 195003 (2006).

[23] Park, H. K. Donné, A. J. H., Luhmann Jr., N. C., Classen, I. G. H. *et al.*, *Phys. Rev. Lett.* **96**, 195004 (2006).

[24] Classen, I. G. H, Westerhof, E., Domier, C. W., Donné, A. J. H. *et al.*, *Phys. Rev. Lett.* **98**, 035001 (2007).

[25] Horton, W., *Turbulent Transport in Magnetized Plasmas*, World Scientific, London, 2012.

[26] Bendat, J. S. and Piersol, A. G., *Random Data: Analysis and Measurement Procedures*, 3rd ed., Wiley, New York, 2000.

[27] Sattler, S., Hartfuss, H. J. and W7-AS Team, *Phys. Rev. Lett.* **72**, 653 (1994).

[28] Cima, G., Bravenec, R. V., Wootton, A. J., Rempel, T. D. *et al.*, *Phys. Plasmas* **2**, 720 (1995).

[29] Watts, C., In Y., Heard, J. *et al.*, *Nucl. Fusion* **44**, 987 (2004).

[30] Udintsev, V. S., Goniche, M., Giruzzi, G. *et al.*, *Plasma Phys. Control. Fusion* **48**, L33 (2006).

[31] Schmitz, L., White, A. E., Carter, T. A. *et al.*, *Phys. Rev. Lett.* **100**, 035002 (2008).

[32] White, A. E., Schmitz, L., McKee, G. R. *et al.*, *Phys. Plasmas* **15**, 056116 (2008).

[33] Watts, C., *Fusion Sci. Technol.* **52**, 176 (2007).

[34] Mazzucato, E. and Nazikian R., *Phys. Rev. Lett.* **71**, 1840 (1993).

[35] Fonck, R., Cosby, G., Durst, R. D. *et al.*, *Phys. Rev. Lett.* **70**, 3736 (1993).

[36] Zakharov, L. E., Gorelenkov, N. N., White, R. B. *et al.*, *Fusion Eng. Design* **72**, 149 (2004).

[37] Coppi, B. and Zhou, T., *Phys. Plasmas* **19**, 012302 (2012).

INDEX